Time / No Time
The paradox of poetry and physics

Seán Haldane

PARMENIDES

Copyright © 2013 by Seán Haldane

All rights reserved.
Published in 2013 by Parmenides Books,
www.parmenidesbooks.ie, an imprint of Rún Press,
www.runepress.ie, Cork, Ireland.

This book is sold subject to the conditions that it shall not, by way of trade or otherwise, be lent, resold, hired out or otherwise circulated without the publisher's prior consent in any form of binding or cover other than that in which it is published and without a similar condition being imposed on the subsequent purchaser.

Printed and bound in Great Britain by
TJ International Ltd, Padstow, Cornwall.

Cover design by Bite Design, Cork.

The cover image is Hexagram 61 from the I CHING:
wind over lake – 'Inner Truth'.

ISBN: 978-0-9574669-0-6

Dedicated to my father

Desmond O'Reilly Fitzgerald Haldane (1910–1992)

'Think, boy, think!'

CONTENTS

INTRODUCTION   1

1   PARMENIDES AND THE GODDESS   13
Peri Physeos. Parmenides and Plato. The paradox of motion. Idealism. Parmenides, space and time. Parmenides, monism and dualism. Parmenides the poet. The mother of all.

2   THREE UNIVERSES   27
The time universe: Stephen Hawking. The multiverse: David Deutsch. The shape universe: Julian Barbour. Theories of everything. Free will.

3   MORE-THAN-COINCIDENCE   51
Miss Love. Coincidence and more-than-coincidence. The law of series. Precognition series. The subjective experience in series. Event series. Precognition series and event series. Emotion in series.

4   SERIES AND TIME   73
Event clusters. Series against time. Precognition and time. Series and cause. Acausality.

5   AWARENESS OF TIME   85
Definitions of Time. Pulsation and Pulse-waves. Human Pulsation. Consciousness. The demarcation of consciousness. Awareness of time. Flow and resistance. We invent time. Living and spatialised time. 'The rhythmic beat of consciousness.'

6   **MIND**   111
Unconscious memory. Memory as mind. Images of the future. Future memory. Past lives. Psychic poaching. The outflow of perception. Where is consciousness?

7   **THE ORIGINS OF POEMS**   133
Inspiration. Poem 1, The Dagd. Poem 2, This and That. Poem 3, Mumm's Champagne. Poem 4, Canadian War Memorial, Green Park. Poetry and the brain. The pre-existence of poems. Poem 5, Black Hill. Where a poem comes from.

8   **PERSISTENCE IN THE COSMIC OCEAN**   155
Seeing time. Orientation. Continuum and discontinuity. The whole and the parts. Dualities. Pulse waves in the cosmic ocean. The duality of the universe. Sex. The time sense. The ocean everywhere. Language, flux and meaning. Change as time. Eternity in an hour. The Unshaking Heart of Truth.

BIBLIOGRAPHY   187
Sources referred to in text.

ACKNOWLEDGMENTS

I have written and re-written *Time / No Time*, under changing titles – a mouse born out of a mountain I sometimes think – on and off for twenty years as I have moved from place to place working in various health services and writing and publishing (or not publishing) other books. I have discussed aspects of this book with a few friends across the years: Adolph Smith, biophysicist (eventually for NASA); Bernd Laska, philosopher of anarchism; Martin Seymour-Smith, poet and radical social thinker, now dead, who always encouraged me to face myself; David Stove, philosopher, also dead, whom I knew only by correspondence but whose work reinforces my belief in what he called 'the inevitability of realism'; Jose Ignacio Xavier, psychiatrist, with whom I have been working on 'neurodynamic psychotherapy'; Jaak Panksepp, neuroscientist of the emotions in both humans and animals; Julian Barbour, theoretical physicist of 'the end of time'. Other friends David Cameron and Rory McTurk have given constructive criticism of the text. None of these people, of course, are responsible for anything I write here. But discussions with them have kept me on an even keel.

I have not discussed this book very much directly with my wife Ghislaine, but she has been my constant intellectual companion and almost all my views have somehow been challenged or refined in discussions with her. The person who has had the strongest influence on my approach to questions of meaning is my father Desmond Haldane to whose memory I dedicate the book. I often hear him say: 'Think, boy, think.'

INTRODUCTION

Poetry and Physics
Poetry is 'making' ('poiein' in Greek) and 'seeing' (in Old Irish the word 'file' means poet and seer). Physics is about 'nature' and 'being' ('physein' in Greek). Their approaches are different: a poem is a one-off, a singular event, whereas physics requires replicability of events. The poet and the physicist see differently, and the physicist's observations are often made indirectly, via instruments. But they see the same nature.

Usually they don't take each other seriously. The 'one off' has no place in physics which by definition (Joe Rosen) consists of a deliberate research programme which looks for repeatable and replicable observations and results. Furthermore each theory in physics must be phrased in such a way that it can be tested and, if 'falsified' (Popper's term) then modified or replaced by a new one. The equivalent in a poem is its being true to the experience of the reader (which is more likely if it was true to the experience of the poet) and it may even modify the reader's experience, his or her way of seeing things. A poem is a vision. And once it is written it cannot be replicated, only copied. The vision of the poet / seer is not incorporated into physics. And since poets see for themselves it is hard to imagine a poem that relies on instruments for its observations. Poets and physicists seem to inhabit separate worlds. But they don't. They share the same world, the same nature. Although the physicist (like his or her historical predecessor the philosopher) looks for generalisable statements, the poet looks for what Blake called 'the minute particular' – again the one-off.

I would like to think that in the late 20th century physics and poetry began to converge: quantum physics explores a universe as strange as that of poetry, and like poetry it puts into question time and logic. But quantum physics seems to be a special case for matter at a very small scale, and its effects are not observable in the human-scale world – either for physicists or poets – unless such one-off phenomena as telepathy and precognition are seen as analogous to the one-off states of sub-atomic particles. And the 'miraculous' (a miracle being by definition a one-off vision) occurrence of poems through what Robert Graves called 'more-than-coincidence' cannot be explained through physics – because it is not testable, replicable etc.

On the other hand, as Rosen has explained, a cosmological theory in physics, although it may have explanatory power and lead to further research, cannot be tested: the universe is too huge and may not even have limits. So physicists and poets are equally free to make untestable statements about such fundamentals as time. As physics since Mach questions the absoluteness of time, and sees it as relative (as poetry often does), perhaps there is some convergence after all. And the cosmological theories of Mach and Einstein are no more testable than those of the first theoretical physicist, Parmenides, who was also a poet.

Parmenides

Over 2,500 years ago the pre-Socratic philosopher Parmenides wrote a poem, of which 161 lines survive as fragments, known as *Peri Physeos*. This is usually translatable as 'On Nature' – 'physis' suggesting not simply what we now call the physical, but what appears as the result of growth and swelling, and also 'being'. Parmenides may have been the first and last person for whom poetry and physics were as one. Soon after his death they split apart, with a shove from Plato. Now with a shove from 20th century quantum physics they may be moving together again. *Peri Physeos* raises perhaps the most basic philosophical and scientific question ever posed: does change exist, or is it only an illusion? Plato, following Parmenides, thought it was an illusion. As Shelley put it, neo-platonically: 'Life is a dome of many coloured glass / Staining the bright radiance of eternity.'

Parmenides, in denying change, was also denying time. He does not mention time. But a universe without change, un-moving, cannot contain time – which is measured by change. Common sense tells us both change and time exist. But the theoretical physicist Julian Barbour in *The End of Time* (2000) separates the two in a proposal that change exists in the shape of the universe but time is an illusion. He describes a timeless universe of pure shape, 'Platonia.' Most physicists do not agree entirely with Barbour, but they would agree that we need to re-think time (and there is an active research programme by a group of physicists who aim to explain anomalies in current physics via Barbour's theory). 'Time's arrow' moving forward into nothingness with history behind it was replaced by Einstein's 'space-time' – which then turned out to be incompatible with quantum physics. Stephen Hawking's *A Brief History of Time* sits on the fence between the Big Bang theory of the universe and the evidence against time. David Deutsch proposes a multiplicity of universes.

I am not a physicist but as a neuropsychologist I work with scientific method: I use statistics and systematic observation to make statements of probability (about for example diagnosis), and I seek further evidence to test my conclusions. Scientific method is effective and widespread because it is not very complex. Most of it consists of counting things which can then in turn be 'counted on' as evidence. As the philosopher David Stove has pointed out, 'Almost any drongo (if an Australianism can be permitted) can do "normal science".' And Stove is not putting science down. But he is not writing about cosmological theories.

At the very least, in this book, I hope to be logical. But I have also written and published books of poems. Certainly the writing of a poem does not follow scientific method. Coleridge wrote that 'A poem is that species of composition, which is opposed to science by proposing for its *immediate* object pleasure, not truth'. But he followed this with: '*Good sense* is the *Body* of poetic genius' and 'Poetry must be more than Good Sense or it is not poetry; but it dare not be less or discrepant.'

Scientific method at the turn of the 21st century has been heavily influenced by Karl Popper who neatly refined the old distinction between induction (reasoning from accumulated observations) and theory (proposing an explanation that may be supported or refuted by observation, with refutation being the more powerful since one observation of a black swan can disprove for ever the 'fact' that all swans are white). Popper described induction as the 'bucket method' of collecting data and then claiming erroneously that the data, rather than theory, suggested a conclusion. He favoured the 'hypothetical-deductive' method in which a theory, or even a conjecture, is advanced in a way that makes it testable ('falsifiable') by observation (evidence). But scientists' accounts of their own theory-making do not always correspond to this ideal. They often state that their theory began with an observation, or when their theory is refuted by a counter-observation they seek further observations which refute the deduction from the counter-observation.

The word 'theory' itself means in Greek ('theoreia') a way of seeing or looking at something. In itself the word betrays a connection, not an opposition, of idea and observation. Some theories are experienced by the person who formulates them as emerging from the observation or as part of it. The *way of seeing* defines what is seen and what is thought about it. Henri Borcroft has explored this in the case of Goethe whose way of seeing colour does not refute Newton's but is nevertheless valid. Borcroft sees this as 'the multiplicity of unity', which is too abstract for me, but it amounts to seeing the parts and the whole at once. The experiential choice is *not* for all observers the black and white alternative proposed by the Gestalt psychologists whose trick drawings show a black vase *or* two white faces – never both at once. Gestalt is claimed to originate with Goethe's ideas, but in my experience (and I think Goethe's) of 'seeing' many levels of meaning *at once* when writing a poem, I find it is possible to see (or experience with other senses) the parts and the whole simultaneously. Only in this way can the complexity of life be understood. Both pure inductivism and pure theory are ways of reducing life to something testable – of reducing, to return to Popper, the 'truth target.' In fact they inter-react, and this is consistent with modern physics where Newtonian mechanisms of action and reaction are replaced by interaction and relation – of the observer and the observed, for example, in quantum physics.

Poetry is *inclusive* of many levels of meaning, among them rational meaning, but it is also supra-rational. Traditionally science and logical argument must be *exclusive* of distractions and stick to a single line of thought. Conclusions (even in the form of probability statements) are more accurate if the truth target is relatively small. But the complexity of modern physics, particularly the discontinuities of quantum physics, has changed this. Whether or not physics and poetry are converging, both must accept paradox.

As I read Parmenides, he does not claim that change does not exist. He states the paradox that it cannot logically exist yet it appears to exist. This kind of paradox is the stuff of poetry. Since the arrival of quantum physics it is the stuff of physics too.

I must admit that many of my observations in this book date back to the decade of the 1970s, which the witty journalist Francis Wheen has described as 'strange days indeed' and (being 'the last pre-digital decade') the origin of an age of Mumbo Jumbo. For the record, I do *not* 'believe in' spiritualism, UFOs, spoon-bending, life energy, the Bermuda Triangle, or deconstructionism. But the decade's craziness in the absence of real innovation (after all, the moon landings had taken place in 1969, and DNA had been discovered in the 1950s) did give a general permission to explore 'anything that goes'. I do, as a poet, leave a space for miracles, and I'm willing to look at anything. But I don't believe in very much, being sceptical. And science is a matter of evidence, not belief. (Or should be: the ruling by an English court in 2009 that 'belief' in global warming has the status of religion is ominous, no matter which side of the debate you are on.) As this book shows, I think there is evidence for precognition, action at a distance or without cause, and fields of consciousness. But these may be incidentals in a timeless universe where we paradoxically experience time.

In this book I concentrate on observations, my own and those of others, which I have replicated. I discuss current cosmological theories and attempt to reconcile them with the observations. I end up with not so much a theory as a description of reality which I invite you to consider and perhaps share. I agree with Stove that 'realism is inevitable,' and I am no idealist.

More-than-coincidence

The non-existence of objective time has been creeping up on modern physics, particularly since the discoveries of relativity and the quantum physics which made a huge break-through in the first 40 years of the 20th century and turned the received world upside down. But this breakthrough was mainly suppressed in the remaining 60 years of the century. Classical physics went on mainly as if nothing had happened. In the 1950s at a school in Northern Ireland known for its scientific excellence, physics was taught to me without any reference whatsoever to the quantum theory which had emerged over 30 years before – while down the road at the Dublin Institute of Advanced Studies the great Erwin Schroedinger who had discovered the quantum wave function was working in quiet obscurity, making waves only in his rather scandalous private life. (He had been rejected when he applied for a post at Oxford because he had two wives, but puritanical old Dublin accepted him.). Although the atom bombs had exploded 12 years earlier I was not taught about Einstein or relativity either. At least now there is an active debate on space-time, spawning academic articles and popular books, and even if many physicists hold to a classical 'arrow of time' view, this is looking more and more like a rear-guard action.

Modern physics is caught in a split. Quantum physics, where particles can disappear and reappear in different places, and 'action at a distance' can occur (even, according to its most radical exponents, changing the *past*) is simply not consistent with the 'classical physics' of space-time and the Big Bang. So there is a stand off: the 'micro' world of particles is given over to quantum physics, and the 'macro' world of the observable universe is given over to classical physics. They never meet. Quantum mechanics is supposedly not detectable in the 'macro' world.

But perhaps it is. We experience something like quantum discontinuities (though erratically and sporadically, like the micro events of quantum physics) in what we call telepathy, precognition, and other 'paranormal' events. Action at a distance at the macro level is staring us in the face. We – like the experimenter in quantum physics – seem to create at least part of the reality we experience and that also creates at least part of us. At particular moments this kind of interaction shows itself in paranormal events. Since I do not want to sign up uncritically to a belief in the paranormal (parapsychology, telepathy, ESP, psychokinesis etc.), I use the poet Graves's term 'more-than-coincidence'. As will be exemplified in this book, it occurs all around us every day.

The scientific approach to more-than-coincidence is to cast it beyond the pale, with poetry and religion. It is a rare event when an eminent physicist such as Freeman Dyson steps out of line in the *New York Review of Books* with an essay 'Debunked! ESP, Telekinesis, Other Pseudoscience'. Although necessarily cautious about 'pure speculation', Dyson invokes Niels Bohr's principle of 'complementarity' – 'that two descriptions of nature may both be valid but cannot be observed simultaneously'. Dyson concludes about the 'paranormal': 'I find it plausible that a world of mental phenomena should exist, too fluid and evanescent to be grasped with the cumbersome tools of science' (presumably meaning the requirements of repeatable and replicable observations).

But although more-than-coincidence contains mental events it is not an entirely mental phenomenon. It includes observable events. For example Rupert Sheldrake's theory of 'morphic fields' to explain telepathy is accompanied by rigorous observations of the *behaviour* (not the unobservable thoughts) of people and animals which suggests that they communicate telepathically.

In this book I set out various procedures for the observation and recording of precognitions and 'series'. I also give examples of how poems appear to have 'the last word' in certain clusters of events. Poems express more-than-coincidence and their forms spring from the duality between biological pulsation – heartbeats and breathing – and universal pulse-waves. Here I am adapting an idea from the controversial psychiatrist and experimentalist in biophysics, Wilhelm Reich who defined life in terms of pulsation but assumed pulsation occurred throughout the universe – e.g. in 'pulsating' stars and the aurora borealis. But there is an easily observable difference between pulsation (as in the unequal phase breathing in and out or heartbeat of an animal or human) and oscillation (as in the equal phase swinging of a pendulum, the vibration of a molecule or the rippling of the aurora borealis): it is the difference between the living and the non-living.

Just as through our experience of pulsation we know time, through our experience of more-than-coincidence we know timelessness.

Physics and poetry
I examine the three universes of Hawking, Deutsch and Barbour: the Time Universe, the Multiverse, and the Shape Universe. (The first and third of these terms are my own abbreviations). All these leave consciousness (or more simply 'awareness'), either as an illusion or a 'mystery'. I propose that pulsation and more-than-coincidence enable the persistence of life against time in 'the cosmic ocean', the space-time continuum. Pulsation enables consciousness which creates time in our minds so that we feel distinct from the continuum – like jellyfish in the sea.

The cosmic ocean is both full of motion and motionless, depending on perspective. Meanwhile we get on with our lives. As Robert Graves put it in 'Midway':

> Nothing that we do
> Concerns the infinities of either scale.
> Clocks tick with our consent to our time-tables,
> Trains run between our buffers. Time and Space
> Amuse us merely with their rough-house turn,
> Their hard head-on collision in the tunnel.

Many modern physicists turn their backs on poetry as a matter of course, but also on other sciences – psychology, neuroscience, and biology – even when these emulate classical physics by being as mechanistic as possible. As Julian Barbour sees it, 21st century physics will become more biological (to include, for example, consciousness, which he sets aside as a 'mystery' in his own work). Unless biology lives up to its name and studies life – not microtomed slices of frozen tissue – it cannot contribute to a study of time. Time, the big theme of modern physics, is inseparable from life.

If mental phenomena are outside the pale of science, what Parmenides called 'double-headedness' looms. Science will take care of the physical body and the mind will be left for mumbo-jumbo or religion or 'imaginative literature.' Mental phenomena then become like God, early in the 20th century, elbowed aside as unknowable, while the physicists get on with the 'theory of everything'.

But since about 1920 physics itself has had to live with a 'double headed' split rather like the famous split in schizophrenia between perception and sensation – the split between classical physics and quantum physics. Dyson's remark about the paranormal could be applied in his own science, where the 'cumbersome tools' of classical physics cannot be applied to quantum physics. The philosopher Colin McGinn notes 'the causal chaos that surrounds quantum theory' and adds that 'the world may not be as well-behaved causally as we tend to think.' Furthermore, 'Science is apt to be speculative; it is not in general some kind of simple registration of the objective facts. And there is a very specific reason for this: the reliance on induction and abduction as ways of going beyond the data… The great prestige of science should not blind us to the very real epistemological concerns it raises – concerns that were quite apparent to philosophers of science during its infancy.'

I have no doubt that many readers will be able to experience the phenomenon of 'more-than- coincidence' if they turn their attention to the details of everyday life and look out for it. Without a doubt the phenomenon exists. The questions that arise about whether it is meaningful or simply random appear to be unresolvable in terms of the statistics of chance. The range of events under consideration is too huge – cosmological, in fact. The laws of chance so far can only be established for extremely simple events such as the tossing of a coin, and even then prediction of what the next toss will bring is random. Again readers can use their own judgement about whether the 'event clusters' and patterns of more-than-coincidence are meaningful or random.

Pulsation and oscillation

I propose that pulsation is the defining characteristic and measure of life. We are aware of time (or invent it) because of the distinction between our biological pulsation and the various oscillations (vibrations, waves and rotations) that surround us in the non-living world. The distinction between pulsation and oscillation is not made in dictionaries where they are usually defined as identical: 'alternating expansion and contraction'. But think of the oscillation of a plucked guitar string or of the sun (which expands and contracts in a regular 2 ½ hour cycle) and the pulsation of a jellyfish, or of our breathing or our heart. Oscillation and pulsation are observably different.

Oscillation is 'equal phase': the expansion and contraction of the sun, or the twanging up and down of the guitar string, are as regular as the swinging of a pendulum. The length (or time) of the expansion equals the length (or time) of the contraction, although as the sun cools and the tension of the guitar string diminishes, the amplitude of the equal phase expansion / contraction phase diminishes.

Pulsation is 'unequal phase': the expansion of the jellyfish's membrane takes about half as long as the contraction which propels it along, just as with the heart's expansion and its contraction, and with breathing in and out.

We are surrounded by regular oscillations or pulse-waves (of tides, of electromagnetic waves, of stars seen twinkling through the oscillating atmosphere, of pendulums) and we pulsate for as long as we live. Pulsation and oscillation are 'out of synch'. We perceive the endless vibrations / pulse-waves in which we live as a continuum. But we are ourselves, because of our pulsation, discontinuous. From our position of discontinuity we are aware of our own 'time' as against the surrounding continuum of apparent time events in motion. The paradox is that modern physics shows that the surrounding continuum is timeless. All that motion is relative, not absolute.

Bergson described in the early 20th century how we 'spatialise' time. Barbour is the ultimate spatialiser. He proposes that time is simply length in a universe of pure shape. In either case oscillation and pulsation can be described as shapes. They will still be distinct. The distinction does not immediately raise the question of the existence of time. But it turns out to have explanatory power. It brings a new perspective to such questions as the demarcation of life from death, dualistic versus monistic philosophy, the distinctions between health and illness, emotion and thought, consciousness and the lack of it, voluntary and spontaneous activity, and how we experience time in what may be a timeless universe.

Pulsation is a shade suspect to those who know of the work of Wilhelm Reich, the neurologist and psychoanalyst who saw it as part of 'the function of the orgasm' and (almost certainly wrongly) as an expression of 'life energy'. I spent some years studying Reich's work, while practising his method of intensive psychotherapy which involves direct work with the emotions and bodily movement. The theories Reich built hastily on his observations are often wrong, and he confused pulsation and oscillation, but the observations themselves are often valid. In this book I discuss Reich and other fringe scientists – or perhaps they could be called scientists fallen from grace – in particular Paul Kammerer, and J W Dunne. It is not that I have an affinity for fringe science. Reich and Kammerer followed their ideas to the edge of madness, and undermined them with hasty and grandiose claims. Having worked for many years in so called 'mental health' (actually mental illness) services I find madness both tragic and monotonous. But keeping in mind the Australian philosopher David Stove's scathing essay 'What is Wrong with our Thought?' in which Plato, Plotinus, Berkeley, Kant, and Foucault get their come-uppance for the *obvious* logical flaws in their theories, I am not sure that Reich and Kammerer are much less mad than various 'great' (or grandiose) thinkers. Both had sound credentials as scientists, were conscientious and even brilliant experimentalists, and made observations worth considering – which perhaps their oddities and immunity from conventionality enabled them to make.

I had a chat with Stephen Hawking about poetry and science when I met him at a college reunion dinner in 1999. (I had known him slightly 37 years before – but in those days at Oxford it was considered bad form to talk about one's own subject, so we probably talked about rowing or beer.) Now we had been talking about a poet we both knew, and I asked him whether he always thought scientifically or did he think poetically too. He wrote on his screen: 'Hard to distinguish. What is the difference between them? Is there a test?' I said I thought there was little difference between good science and good poetry, but a million miles of difference between bad science and bad poetry. His eyes showed amusement. Later in the evening I came back to him with: 'Here's a possible test of good poetry and good science. They both make your hair stand on end.' To which he wrote: 'What about an electric shock?'

This was a fairly typical piece of Hawking mischief. I remembered from years before his tendency to deflect any serious remark with a joke. But he may have been as serious as I was. An inspired poem and an inspired scientific insight both have the power to shock. Both state rational truth but both have begun in sudden revelation or intuition, an interruption in ordinary thought. And I think science can only advance, and poetry can only exercise what Thomas Hardy called its 'sustaining power', if each is open to the thinking of the other. The other side to the death of objective time is the subjective experience of continuing to live it. The interface between the objective and the subjective is surely a right point of concentration for both science and poetry.

I am not arguing that physicists 'should' read poetry or that poets 'should' study physics. But I think there are more points of contact between the two than is usually acknowledged. And one can inform the other. Various ideas from physics turn up in poems. And some physicists – Mae-Wan Ho and Julian Barbour for example – quote poems which support their view of reality.

1

PARMENIDES AND THE GODDESS

Peri Physeos
Fragment 1
>*The mares that carry me as far as my heart desires...put me on the road of a goddess who conveys to all towns the man who knows...*

A young man is in a chariot, its axle glowing with heat and making a whistling sound, pulled by straining mares, and escorted by maidens who have pushed back the veils from their heads. They have left the house of the night to go into the light. Ahead are the gates of night and day – shining brightly and surrounded by stone. The maidens persuade the keeper of the gates, the avenging female figure of justice, to push back the bolt and let them through onto a broad way. They are received by a goddess who takes the young man's right hand:
'Welcome! It was not an evil fate that sent you along this road, far from the beaten track of men, but right and justice. Now you must learn everything, both the unshaking heart of persuasive truth and the beliefs of mortals, in which there is no true trust...'

Fragment 2
>*'Come now. I shall tell you the story, and you shall tell it to others, of the only paths for thinking. The one – **that it is and it is impossible for it not to be** – is the path of persuasion, for it accompanies truth. The other – **that it is not and it is right for it not to be** – is a path from which no tidings ever come, for you could not know what is not... nor could you convey it.'*

Fragment 3
>*'...For the same thing is there for thinking and for being.'*

Fragment 4
> '*Gaze even on absent things with a present mind, and do so steadily...*'

Fragment 5
> '*It is the same to me where I start from as there I shall return.*'

Fragment 6
> '*...What is is. For it can be, but nothing cannot...*'

The goddess urges the young man to think only about this truth, and constrains him from the other path she has mentioned, the path of opinion, in which people, 'double-headed', going back and forth, allow themselves to think of 'nothing' and consequently wander adrift, tossed about, confused between what is and what is not (nothing), turning back upon themselves.

Fragment 7
> '*For this will never be shown: that things that are not* ***are*** *...Do not let habit do violence to you on the way of experience so that you use unseeing eyes and an echoing tongue...*'

Instead the young man must hold to reason and to the difficult proof she has given (namely that **'What is is. For it can be, but nothing cannot'**).

Fragment 8
> '*Only one story of the way remains: that it is...*'

The goddess provides an argument that existence can only be described in terms of itself – of what is. Any thought of what is not – of what is existing in a universe where there is also 'nothing' – will lead the young man off the way of truth onto the erroneous way of opinion. What *is* is whole in itself, unmoving, perfect, altogether one, cohesive; it is a plenum, a fullness in which there is no nothing. To think of what is not is to make a decision not to think of the truth of what is. There is no growth, no future, no birth – because birth requires that what is born comes out of nothing.

'How could that which is be later on? How could it possibly get to be? For if it got to be, then it is not... Immobile within the bounds of great fetters, it is without start or pause, since coming-to-be and perishing have strayed far and wide and it was true fidelity [to the truth] *that drove them off. The same and remaining in the same, it lies by itself and in this way remains there fast, for mighty necessity holds it in the fetters of a bond which restrains it all around...'*

Mortals have named all sorts of changes – coming to be, perishing, being and not being, changes in place, changes in colour – none of which are true.

'For there neither is nor ever will be any other besides what is, since it was this which fate bound to be whole and unmovable...'

The goddess describes the universe as a well rounded sphere, equal everywhere and from every side, and all the same. She then concludes her description of the truth and returns to the appearances or seemings which lead mortals to two ways of naming things: one right way – naming them as what is, what they are; and one wrong way – naming them in terms of distinguishing characteristics, such as contrasting light with dark, lightness with heaviness, so that each characteristic gives a semblance of things existing by themselves, not in the whole.

Fragment 9

'...all is a whole of light and of obscure night, together, both equal, since nothingness partakes in neither.'

Fragment 12

'The narrower rings are filled with unmixed fire, the next ones with night and an amount of fire shoots out. In the middle of them is the goddess who steers everything. For she rules over deadly birth and mixing together of all things, sending the female to unite with the male and the male to unite with the female...'

Fragment 13
> 'She created Eros first of all the gods.'

Fragments 14 & 15
> [The moon] ' shining in the night, wandering around the earth, a foreign light…always looking towards the sunshine.'

Fragment 16
> 'Mind is present to men according to the temper which it has of the wandering body at each moment. For it is the same thing, the nature of the parts, which thinks in all men [as one] and in each [separately]. For the full is thought.

Fragment 17
> 'When man and woman mingle the seeds of love that spring from their veins, a formative power, maintaining proper proportions, moulds well-formed bodies from this diverse blood. If, when the seed is mingled, the forces clash and do not fuse into one, then cruelly will they plague the offspring with a double-gender. On the right [of the womb] boys, on the left girls.'

The above is a mixture of text and paraphrase of the surviving 161 lines, in fragments, of Parmenides' great poem in Greek which is known as *Peri Physeos*, or *On Nature*. Even the title invokes controversy, since 'physeos' already contained ideas of dynamic movement in the form of growth and swelling in nature – which Parmenides' unmoving plenum rejects. But does it? For one thing, the speaker in the poem is 'the goddess' and not the young man, although it is he who is conveying her message according to her instructions. And for another, although Parmenides is traditionally identified as favouring the unmoving universe, and the goddess states that she does too, in the poem she reveals both sides of the story.

The reading of Parmenides' text is fraught with controversy, about textual readings and such linguistic points as the meaning, for Parmenides, of the words for 'to be' and 'is', which have a huge effect, in turn, on the reading of his philosophy. I know almost no ancient Greek. But I have cobbled together the above version of the text, when I quote the goddess directly, from translations (of the whole text or parts) by Austin, Coxon, Mourelatos, Popper and Rendall. I have gone back to the Greek via textual commentaries when translating key words.

Parmenides and Plato

Parmenides of Elea probably lived from about 515 BC until 450 BC. The fragments called *Peri Physeos* are all we have of his writings. There is a fragment (17, above) surviving only in a translation into Latin, sometimes included in *Peri Physeos* but not necessarily part of it, about how the 'seeds' from man and woman in the womb may sometimes get mixed up and lead to offspring with a 'double gender' – a view which prefigures the modern neuroscientific findings that sometimes a 'female' brain may be found in a male body and vice versa because of disturbed biochemistry in the foetus. Parmenides was no slouch when it came to proto-scientific thinking. His unmoving universe was reborn in the late 20th century in Julian Barbour's 'Platonia' – a timeless universe consistent with quantum theory. 'Platonia' could more accurately, though more clumsily, have been called 'Parmenidonia'. But somehow Plato got in between. His dialogue 'Parmenides' coopts Parmenides' unmoving plenum into Plato's own project of idealism. Plato's 'forms' which underlie the illusory universe of our senses owe everything to Parmenides' plenum of truth which is obscured by the perpetually changing and illusory world of appearances.

Plato's presence in the history of philosophy is like a wall between us and Parmenides. Behind this wall, classical scholars scurry, elucidating Parmenides' texts. And like a banner raised above it, the gist of his philosophy survives as his 'timeless universe', or as a 'pre-Socratic' pole of 'everything is static' in opposition to that of his predecessor Heraclitus, 'everything is flux and nothing is at rest.'

In the 20th century, though, several philosophers and scientists have attempted to get to grips with Parmenides. Einstein's theories imply a block universe of 'space time', and he was humorously nicknamed 'Parmenides' by Karl Popper in a discussion they had about relativity. The first self-conscious philosopher of science, Popper wrote a book on Parmenides as the first philosopher of science... in that Parmenides, although absurdly wrong in Popper's eyes in having no place for time in his universe, was the first to express the need to test theory by experience – as the goddess urges, to keep the gaze on what is absent [from the senses], not on the ever changing or seemingly changing illusions of what is not. He was also the first philosopher to offer a sustained logical argument. Popper summarises this in prose :

> Only what is, is.
> The nothing cannot be.
> There is no empty space.
> The world is full.
> Since the world is full there is no room for motion – and thus for change (which is a kind of motion).
> Motion and change are impossible.

It has become a cliché that Parmenides' universe is timeless. But it is worth pointing out (as the textual analyst Mourelatos confirms) that he does not use any words for time in Peri Physeos, nor even indirectly refer to it. Nor does the goddess's argument, or Popper's redaction of it, draw attention to time. Timelessness is simply a logical consequence of an unmoving plenum: a whole with no birth, no growth, no past, no future.

The paradox of motion
The goddess shifts between her condemnation of the 'what is not' path of changeability and a vivid description of it which is just as real as her preceding description of the 'what is' path of unchanging truth. The young man is a traditional traveller or shaman on the quest for knowledge whose spirit ('thymos' – to the Greeks the seat of longing and feeling, in the chest) is transported to the presence of a divinity who reveals the truth. He has been hurtled along the road on a chariot with straining mares and its axle glowing with heat and whistling as the wheels turn, as maidens pushing back their veils escort him on his journey: the whole account is alive with movement. The poem is replete with the paradox that although the universe is unmoving and unchanging, it appears to be moving and changing. According to Plato, one of Parmenides' students was Zeno – who might be called the official father of the paradox, for his conundrum of

the hare and the tortoise, and for pointing out that since an arrow at any instant occupies a particular place, its motion must be an illusion.

Aristotle presented Zeno as believing there was no such thing as motion (as the goddess claims in the first half of *Peri Physeos*) but it may be more accurate to say that Zeno was calling into question our usual assumptions about movement. He may have been the first to suggest that movement was a matter of perspective – something that, as Julian Barbour has documented in his monumental *The Discovery of Dynamics*, took over two thousand years in the history of physics to become understood. Zeno came into his own with the philosophy of Henri Bergson at the beginning of the 20th century. Bergson's distinction between 'lived time' (temps vécu) and 'spatialised time' (temps spatiale) goes back to Zeno – and to Parmenides.

The Australian philosopher David Stove remarks indignantly that Parmenides *travelled* widely to teach the idea that *nothing moves*, and must surely have known that *he* moved, so his teachings were nonsense. But this received view of Parmenides is not consistent with the text of *Peri Physeos* which is, again, about the paradox that things *do* move even while part of a solid and eternal plenum. In the poem itself the young man *travels* from the house of darkness into the light of the goddess's teaching that nothing moves and that all is a whole of light and obscure night.

Idealism

If it is acknowledged that Parmenides is the grandfather, as it were, of Plato's idealism, then this is a line which has gone a long way. What the idealism-bashing Stove calls the 'Plato Cult' has not only (via Aristotle, Plotinus and a hundred others) provided the backbone to Christian ideas of eternity, anti-materialism, and the duality of spirit and flesh. It has held back the growth of science, not to speak of common sense, by turning people's official thoughts (not their naturally resistant private ones) to 'higher' things than mere facts, to vast generalisations about 'universals' rather than what William Blake called 'the minute particular'. By the mid 18th century, neo-Platonism had led to Bishop Berkeley's contention that the universe was just a dream in the mind of God. The late eighteenth century Enlightenment, for all the faults of extreme rationalism and positivism, at least threw up thinkers such as David Hume who brought us back to earth under what he called the 'calm sunshine of the mind', in contrast to what he called the 'sick men's dreams' of idealism. Modern science, as a social discipline (though of course there were 'scientific' thinkers before 'science' as such became a term, in the 19th century) is based in the Enlightenment. Popper has referred to the 'pre-Socratic Enlightenment' in tribute to the serious attempts by Parmenides and his near contemporaries to discover the nature of reality. But Hegel's version of idealism renewed the neo-Platonic dream and saturated the nineteenth century (see Stove's double essay, 'Idealism: A Victorian Horror Story'). It is still alive in the 'Intellectual Impostures' of Foucault and others. But this fuzziness is no longer recognisable as having any origin in Parmenides whose thinking has clear edges, as it were, like the drawings of Blake.

Parmenides, space and time

One line back to Parmenides is from the late Renaissance when in the 16th and 17th centuries astronomers began to overturn the medieval and ultimately Aristotelian / Ptolemaic universe of perfect spheres within spheres, the earth at the centre. They did so by studying change – so much so that John Donne, who agonised about Kepler, whom he read in manuscript, lamented that

> ...new Philosophy calls all in doubt,
> The element of fire is quite put out;
> The Sun is lost, and th'earth, and no man's wit
> Can well direct him where to look for it.

By the mid 17th century the Royal Society was established in England, and the intellectual climate was favourable to innovative, even eccentric ideas such as those of Leibniz who wrote: 'Space is nothing but the order of co-existent objects; time is nothing but the order of successive events.' A modern commentator on Leibniz summarises:

Space and time are just ways (metaphysically illegitimate ways) of perceiving certain virtual relations between substances. They are "phenomena"; that is, in an important sense illusions – although they are illusions that are well founded upon the initial properties of substances. Thus 'illusion' and 'science' are fully compatible.

These are the two ways expounded by the goddess in Peri Physeos.

Parmenides, monism and dualism

Then there is the line to Parmenides as the first philosopher of science, as explicated by Karl Popper (in a book which came out posthumously). There is something comic in Popper discovering in a 5th century BC poem his own theories of trial and error, 'conjecture and refutation', as the benchmarks of true science. Popper even states that when the goddess talks about truth she means 'truthlikeness' – a phrase of Popper's which she presumably anticipates. But Parmenides probably deserves Popper's tribute: 'Parmenides' theory may be described as the first hypothetico-deductive theory of the world.' Popper goes on to say that Parmenides' successors 'the atomists took it as such; and they asserted that it was refuted by experience, since motion does exist.' Popper summarises this counter-argument:

> *There is movement.*
> We know this from experience.
> Thus: The world is *not* full;
> there *is* empty space.
> The nothing, the void, does exist.
> Thus: The world consists of the existing,
> of the hard and full *and* of the void:
> Thus: The world consists of *'atoms and the void'*.

In other words, as Popper concludes, 'the world is dualistic.'

So Parmenides has served the first ball in an argument (monism versus dualism) to which the above is the return and which has not yet been settled. But this cuts Parmenides' poem in two. Yes, the goddess is a monist as she expounds the plenum. But *Peri Physeos* expresses both sides of the argument: there is as much of Heraclitus as Parmenides in it. And since Parmenides wrote the poem he can be taken literally. The 'Parmenides' of tradition is the voice of the goddess in the poem. But what about the Parmenides rushing on his chariot through a universe in flux? If we step back and read *Peri Physeos* without prejudice it is clear that Parmenides represents both sides of the argument. The poem is not monistic but dualistic – though not in the atomist sense of matter versus the void, which along with Platonism was a precursor of the later matter versus mind dualism. We are so used to this dualism which has endured for millennia that we tend to think it is the only one. But Parmenides' dualism is between two states of matter: the moving and the unmoving.

Popper cites Julian Barbour:

> Then there is the problem of reconciling quantum mechanics with Einstein's theory of gravitation. Quantization of that theory in the case of a spatially closed universe by the standard methods leads to a 'wave function of the universe' that appears to be completely static. Nothing happens at all – there is a complete Parmenidean stasis, in flagrant contradiction of the evidence of our senses.

Parmenides the poet

How much was Parmenides simply writing in verse, at the end of a period when this gave authority to any utterance, and how much was he a poet? His predecessor Anaxamander had also expressed his philosophy in Homeric hexameters. The difference between verse and poetry is usually decided by the reader's sense of what he or she defines as poetry (including such aspects as emotion, compelled rhythm, vision, 'inspiration'), and this is hard to establish for ancient Greek. The Greek scholar Mourelatos has claimed that Parmenides was not an inspired poet. But he himself quotes another scholar to the effect that the line about the moon is one of the most beautiful in Greek poetry, and devotes a page to a subtle exposition of its depths. It compresses several layers of meaning into just six words, with an assonance of 'O' sounds suggesting the round shape of the moon [one of the two 'o' vowels in Greek does not look like an 'o', but the mouth has to be rounded in saying it], and a pun in the last two words in which *allotrion phos*, 'a light not its own', is almost identical to the Homeric *allotrios phós* 'an alien man' or 'stranger'. Certain phrases in the poem would stand out in any language, and although as Robert Frost said, 'poetry is what gets lost in translation', they come through strikingly no matter how translated. For example (fragment 1) *'the unshaking heart of persuasive truth'*, even if *atremes etor* is translated as 'unmoving heart' or 'steady heart' – or as is possible in another interpretation of *etor* as meaning 'guts' or 'innards', *'the unstirred guts of persuasive truth.'*. Or the goddess's lines (fragment 4) *Gaze even on absent things with a present mind'*, and (fragment 5) *It is the same to me where I start from as there I shall return.'*

Besides which, the whole poem consists of a *vision* – such as Old Irish writings describe as a *fís* or an *aisling* (in which the vision is of a fairy woman or goddess) – which attests to its inspiration. And finally the paradox which is at its core is characteristic of the poetic way of thought – much more than of the philosophical or the scientific ways.

The paradox is simply that Parmenides – or the goddess who speaks the poem – is clearly on both sides of an irreconcilable question. This does not seem to have been apparent to Plato, or he chose to ignore it. He presents and modifies only one side of Parmenides' position. But the paradox was apparent to others at or close to the time. The historian Plutarch notes drily that 'Parmenides... both laid claim to his opinions, and at the same time took the opposite standpoint.'

Yes, the universe is a whole in which every part is connected, the concept of 'nothing' or 'no thing' is an illusion, and therefore there is no past and no future, just connection. But yes, we live in a world of changing impressions, of 'seemings' in which we are born and live and die, and most crucially we love – erotically, as men and women – and perpetuate change by having children. The goddess tells the young man to keep his gaze on the unchanging truth – while the universe changes all around him. She describes, vividly, this change. The paradox is that although she favours the unchanging world of persuasive truth and urges the young man to shun the changing world of opinion, each world is one way of truth.

Poetry always speaks like this. (Unless it is in the hands of time-servers and sycophants who express in gracious verse the conventional wisdom of their day). In English poetry, think of John Donne, with one eye on eternity and the other on his mortal woman, or William Blake, with one eye on innocence and the other on experience, or Thomas Hardy who said that all his poems were about 'seemings' – the world of appearances the goddess describes in fragment 8 – but who wrote of 'the sustaining power of poetry'. This is not a comfortable way of seeing the world, perhaps, but it is the poet's way, and it speaks to all of us, because we are all caught in this paradox.

Goethe founded a whole system on it.

Plato famously banished poets from his ideal Republic, to be run by philosophers. He had taken only one side of Parmenides' argument, that 'the same thing is there for thinking and for being', and the emphasis by the goddess on keeping one's gaze on what is changeless. But he had no liking for poetry, for paradox, or for the minds of women – who traditionally inspire male poets. (Sappho, who lived before Plato, inspired herself, as perhaps women poets must do).

The mother of all

Parmenides is writing in a tradition which is almost extinct. Poetry is now, as the Scottish Gaelic poet Sorley MacLean wrote in the 1930s, 'a dying tongue' (in Gaelic 'cainnt bhasmhoir'). But Parmenides' poetry has its origins in 'Old Europe', the neolithic and bronze ages when there seems to have been a common imagery of religion focused on women, on fertility, on carvings or paintings of zig-zags, spirals, and animal forms in pottery or on rock or wood, in which the forms of natural change recur, as in nature itself, and all parts of nature seem to be interconnected. The archaeologist Marija Gimbutas has explored this in *The Gods and Goddesses of Old Europe*. The poet Robert Graves in *The White Goddess* interpreted the myths of how Old Europe was succeeded by a more patriarchal way of thought – typified first by Socrates and Plato – where the ancient imagery gave way to rationalism. Parmenides, whom Graves does not discuss, stands between Old Europe and the New Europe of Platonic philosophy and eventually of Christianity. The duality of his poem, in which the world of change co-exists with another world of changelessness, would have been familiar to the Celts of his time who believed (as later Old Irish texts show) that the everyday world is paralleled by a timeless Otherworld.

In *Peri Physeos*, there is only one male – the narrator. The other characters are all female: from the maidens who escort the chariot (and for that matter the mares who pull it – who would have been sacred to the goddess Demeter), to the nameless goddess herself. Yet various goddesses are named: *Dikhe* – the vengeful justice at the gates of night and day; *Ananke* – Constraint, or Necessity; and *Moira* – Fate. Commentators tend to agree that all these are in fact names for the nameless goddess, a 'polymorph deity' as Mourelatos puts it, who teaches the young man the ways of truth.

Parmenides states the paradox of motion – as set out later by his student Zeno: if you *measure* it, it does not exist – especially if you consider the universe in a calm and *measured* way. But we do not. We see the universe in a whirl – most of the time. Parmenides says, in effect: Pause and see things as they are. Let go of the whirl.

This exhortation may have affinities with the older pre-literate traditions of Old Europe as reconstructed by Gimbutas. It clearly has affinities with some Oriental traditions roughly contemporary to Parmenides, such as Buddhism with its injunction to get off the wheel of becoming. And the change / no change paradox is a recurring theme in the world's longest tradition of poetry, in China. Su Shi in *The Poetic Exposition on Red Cliff*, written about a thousand years ago but referring back to the poetry of a thousand years earlier still, writes: 'If you think of it from the point of view of changing, then Heaven and Earth have never been able to stay as they are even for the blink of an eye. But if you think of it from the point of view of not changing, then neither the self nor other things ever come to an end.'

The injunction of the goddess to Parmenides to keep his gaze on the unchanging universe became Plato's eternal world of immobile 'forms'. But these 'universals' which lie behind 2,500 years of idealism are already an abstraction from Parmenides' concrete, unchanging 'physeos.'

Perhaps echoing Parmenides' goddess, Ernst Mach in the 19th century saw the universe as entirely interdependent, with no part of it, *not even motion*, separable from the totality. The universe was the *mater omnium* – the mother of all.

2

THREE UNIVERSES

The Time Universe: Stephen Hawking
The current prevailing theory of time is that summed up in Stephen Hawking's *A Brief History of Time*. The sub-title says it all: 'From the Big Bang to Black Holes.' The first few chapters clarify.

Chapter 1, 'Our Picture of the Universe' is a brief survey of cosmology from Aristotle (No mention of Parmenides or even Plato in this chapter or in the book) to Edwin Hubble and the 'red shift' of light from distant stars, a sort of cosmic Doppler effect like the descending sound of a car horn as it speeds by, which is the main evidence for an expanding universe. Hawking skips the 19th century physicist Mach (no mention in the book) who, anticipating quantum theory and Heisenberg's Uncertainty Principle, proposed that our theories of reality or the universe are dependent on our sensations. He mentions that Aristotle's universe was 'static', and notes that at the time of the Aristotelian St Augustine 'When most people believed in an essentially static and unchanging universe, the question of whether or not it had a beginning was really one of metaphysics or theology.' In other words the view of the universe as unchanging is not part of science. On the other hand he faces the facts that 'Today scientists describe the world in terms of two basic partial theories – the general theory of relativity and quantum mechanics', the former concerned with large scale observations and the latter with small scale observations. 'Unfortunately, however, these theories are known to be inconsistent with each other – they cannot both be correct.... One of the major endeavours in physics today, and the major theme of this book, is the search for a new theory that will incorporate them both – a quantum theory of gravity.'

Hawking begins Chapter 2, 'Space and Time' with 'Our present ideas about the motion of bodies date back to Galileo and Newton.' He goes on to explicate brilliantly Maxwell's and Einstein's theories of the speed of light.

And by half way through this chapter it is clear that Hawking has bought into the Einstein space time theory: 'An event is something that happens at a particular point in space and at a particular time.'

The middle chapters of the book do not so much explore time as the proposed history of the universe since its beginning with the Big Bang – which is reasonable if the universe is co-terminous with Time, as Hawking appears in these middle chapters to believe. Much of *A Brief History of Time* turns out to be a brief history of the universe as it expands, including a discussion of Black Holes (which although their existence was first suggested by another physicist, John Wheeler, Hawking was first able to explicate) and the 'string theory' of matter which occupies legions of theoretical physicists. But towards the end Hawking tackles his subject head on, and his summary of his arguments is worth quoting at length:

> To summarize, the laws of science do not distinguish between the forward and the backward directions of time. However there are at least three arrows of time that do distinguish the past from the future. They are the thermodynamic arrow, the direction of time in which disorder increases [entropy]; the psychological arrow, the direction of time in which we remember the past and not the future; and the cosmological arrow, the direction of time in which the universe expands rather than contracts. I have shown that the psychological arrow is essentially the same as the thermodynamic arrow, so that the two would always point in the same direction. The no boundary proposal for the universe predicts the existence of a well-defined thermodynamic arrow of time because the universe must start off in a smooth and ordered state. And the reason we observe this thermodynamic arrow to agree with the cosmological arrow is that intelligent beings exist only in the expanding phase. The contracting phase will be unsuitable because it has no strong thermodynamic arrow of time…. The progress of the human race in understanding the universe has established a small corner of order in an increasingly disordered universe.

It is fascinating how Hawking, while setting out an orthodox (for the turn of the century) view in which it is clear that he believes the universe is expanding from the Big Bang and may eventually contract, and that in the current expanding phase the laws of entropy apply, almost mischievously allows some quite unorthodox ideas in. The thermodynamic arrow only agrees with the cosmological arrow because *intelligent beings exist only in the expanding phase.* And these are *a small corner of order in an increasingly disordered universe.*

When *A Brief History of Time* was first published in 1987 and became a best-seller, there were many jokes about how millions of people bought it and never read it. Most of the people I met who had bought it had at least made a good stab at it. But it is fair to say that most readers probably absorbed the more orthodox message – about the Big Bang and the expanding universe and the validity of at least one arrow of time and the unremitting presence of entropy – but did not detect that Hawking had not really fulfilled his initial agenda and provided even a step towards a quantum theory of gravity, and that he had inserted a rather grand role for the way we, human beings, help construct the universe. An almost 'Machian' point, although he did not mention Mach.

Perhaps I am influenced by the fact that I knew Hawking when we were at the same Oxford College, although not especially well as he was two years ahead of me. But I remember him as an essentially mischievous character. He was the cox of one of the college rowing eights, and he participated in a joke 'coxed fours' race, open to all, where the rule (very daring for 1962) was that the cox had to be a woman. I rowed in one four, having brought in a very tiny woman student to be the cox. We were beaten by a four put together by Hawking, who was rowing, and which had a dummy woman for cox. (I hope I remember this right). From another source (Martin Seymour-Smith who tutored him when as a boy he visited Mallorca with his mother, a friend of Beryl Graves), I know that Hawking as a boy had a propensity for throwing stink bombs under tables when people were having dinner. Of course he became tragically immobilised by illness and I can understand he might have a revulsion for any vision of a 'static' universe. But I see in *A Brief History of Time* both the brilliant exponent of what can be called the Time Universe and a sort of 'Till Eulenspiegel' (prank-player) who is honest at the core and cannot stop himself from slipping in some hints at an alternative.

In his most recent book, *The Grand Design*, Hawking adds little to the view of time in *A Brief History*, although he elaborates the currently fashionable inferred 11 dimension universe of 'string theory'. (For a sceptical analysis of this, see Lee Smolin's *What is the Trouble with Physics*). In most modern theoretical phsyics (Hawking, Penrose, Deutsch – see below) time is not absolute but it does not disappear (see Barbour, below). As 'space-time' it is taken to begin with the Big Bang and progress from there. Whether some form of space-time existed before the Big Bang is an unanswerable question. Again recently, Roger Penrose has proposed in *The Cycles of Time* that space-time begins with the Big Bang, progresses to the end of the universe in collapsing black holes and total entropy, then reconstitutes itself in a new Big Bang. Whether absolute or not, in this model 'story' time is linear.

The Multiverse: David Deutsch

Following Hawking, the theoretical physicist David Deutsch has taken the quantum angle on cosmology all the way, with his theory of many universes. His book *The Fabric of Reality* starts harmlessly enough with superb teaching chapters on quantum mechanics. Discussing the various split beam experiments that demonstrate the behaviour of photons, he writes:

> This property of appearing only in lumps of discrete sizes is called *quantization*. An individual lump, such as a photon, is called a *quantum* (plural *quanta)*. Quantum theory gets its name from this property, which it attributes to all measurable physical quantities – not just to things like the amount of light, or the mass of gold, which are quantized because the entities concerned, though apparently continuous, are really made of particles. Even for quantities like distance (between two atoms, say), the notion of a continuous range of possible values turns out to be an idealization. **There are no measurable continuous quantities in physics.**

I have bolded the last key sentence. Quantum physics is about discontinuity. Or perhaps only about discontinuity in measurement. Controversy within quantum physics tends to hinge on whether discontinuity is only apparent or real. For example David Bohm's 'implicate universe' attempts to rehabilitate continuity by an analysis of quantum wave function which is at odds with Deutsch's and most others, and even proposes an explanation for (continuing) consciousness in terms of quantum functions in the brain. This is not generally accepted. However, Deutsch's pursuit of discontinuity to its logical conclusion is not generally accepted either. He proposes that at each instant where a discontinuity occurs, i.e. when the quanta re-arrange themselves from one state to another, a new universe occurs.

Deutsch explains how the micro-world of quantum physics does apply, in spite of what seems common sense, to the macro-world of our experience, because we are composed of micro processes. Discussing Dr Johnson's dismissal of Berkeley's idea that everything existed only in the mind of God by kicking a stone and saying 'I refute it *thus*', Deutsch remarks, 'Dr Johnson did not directly kick the rock, either. A person is a mind, not a body....Dr Johnson's mind, like Galileo's and everyone else's, 'kicked' nerves and was 'kicked back' by nerves and inferred the existence and properties of reality from those interactions alone.' Later Deutsch remarks, 'If something 'kicks back', it exists.' This is a strong support for the idea, discussed earlier, that awareness (consciousness) and indeed life depend on *resistance*. (Death, which we perceive as a void, is the end of resistance.)

I am going to pass quickly by Deutsch's description of computational physics and his idea that as we never experience reality directly, 'Every last scrap of our external experience is of virtual reality.' For me it comes uncomfortably close to saying that 'thoughts are things'. I think, for example, a poem has a reality in itself, but it is not what it describes. But I find all thinking difficult when it becomes metaphysical or philosophical. I would describe myself as quite 'concrete' in my thinking, quite 'literal.' I do not like abstractions. I usually prefer what Blake called 'the minute particular' to a generalization. But I find that once Deutsch leaves virtual reality behind he surpasses me in the concreteness of his thinking. He states that 'the laws of physics do not merely mandate their own comprehensibility in some abstract sense... They imply the physical existence, somewhere in the multiverse, of entities that understand them arbitrarily well.'

In interference experiments tangible or observable particles are clearly influenced by the presence of other particles which are invisible, intangible – 'shadow particles.' 'Thus we have inferred the existence of a seething, prodigiously complicated, hidden world of shadow protons… The only thing in the universe that a shadow proton can be observed to affect is the tangible photon that it accompanies.' Yet (earlier), 'there are many more shadow protons than tangible ones.' The logical inference from the experiments is that 'particles are grouped into parallel universes.'

This idea originated with Hugh Everett 50 years ago and was developed by John Bell. Deutsch has made it very much his own, setting out tenacious arguments in its support. These are concrete because he insists on taking the interference phenomena literally whereas, as he remarks, most other physicists take them 'as if' and argue that 'quantum theory is about the *interaction of the real with the possible*.' And this is the mainstream position, the sort of stand-off as described by Hawking (above) in which the question of how the micro-reality of quantum physics and the macro-reality of classical physics are to be resolved is still open. For Deutsch it is not still open, it is settled. It is for Julian Barbour too, although as we shall see he interprets the 'big picture' from a quantum physics point of view in a somewhat different way from Deutsch.

Meanwhile, in parallel universes there are many versions of David Deutsch. 'Many of those Davids are at this moment writing these very words. Some are putting it better. Others have gone for a cup of tea.'

Deutsch, of course, is a living being. But he seems to have very little interest in this. He does not distinguish virtual reality from consciousness. He does express a sharp critique of the kind of school science teaching that turned me off the subject for some years. 'Thus, by the time I learned biology in school… Life was not considered to be fundamental at all. The very term 'nature study' – meaning biology – had become an anachronism. Fundamentally nature was physics… My classmates and I had to learn by heart a number of "characteristics of living things". These were merely descriptive.' Deutsch goes on to list the usual *respiration, excretion, reproduction, growth, irritability* ('kick back'). Mentioning the Aristotelian view that life is an 'essence', he states, 'We no longer expect there to be any such essence, because we now know that 'animate matter'…is not the basis of life. It is merely one of the effects of life, and the basis of life is molecular. It is the fact that there exist molecules which cause [sic] certain environments to make copies of those molecules. Such molecules are called *replicators*.' He then proceeds inevitably (citing Richard Dawkins) to genes, not as just one element in the life process but as the be all and end all. And 'Genes are in effect computer programs… expressed in a standard language called the *genetic code*.' Deutsch, who elsewhere in *The Fabric of the Universe*, is scathing about 'reductionism', has fallen straight into the reductionist trap of genetic determinism, with its contamination in the work of Dawkins and his followers by an unabashed 'naturalistic fallacy' where the molecules behave purposefully. For a logical demolition job on this view, see the philosopher David Stove's *Darwinian Fairytales*. And modern genetics, as for example in the work of Matt Ridley, denies that genes are 'programmes' or even blueprints: they are 'developmental switches', switching each other on and off according to context, i.e. environment.

There is an indeniably personal slant in any scientific enquiry. Bertrand Russell wrote cynically that in any medical article on the subject of alcohol one could deduce by the end of the first page whether or not the author consumed it. I have mentioned Hawking's understandable objection to a 'static' universe and the mischievous component in his ideas. There is nothing wrong with this. As even Popper concedes in *Objective Knowledge* there is always a subjective presence in experiments. This was also made clear in Heisenberg's uncertainty principle. The scientist's theory may be 'out there', on the table, but he or she has had a hand in it – as the poet has had a hand in a poem. Since Deutsch brings 'David' into his argument he cannot object if his own personality comes into an evaluation of his work. I have never met him and know little about him, but there is a strange lack of feeling for life in his book. The cerebrality of his arguments allows him to push through the logical implications of the quantum interference experiments to the end. But for me his multiverse argument is undermined by his lack of interest in its human consequences. How is one supposed to *deal* with the idea that one exists in millions or billions of versions in so many universes? This does not seem to bother Deutsch enough to lead him to discuss it. But then he has already accepted Dawkins's reductionist argument that life just consists of replication of the genetic code. If we are already, in this world, merely accumulations of genetic replications, why should we not be in millions of worlds? And although Deutsch emphasises that we are 'mind', this mind turns out to be no more animated than the virtual reality programme in a computer. And here his argument founders, because the more is known in neuropsychology about mental processes, the less it is possible that a computer could ever replicate a mind. They are entirely different. This is a larger argument than can be discussed here, but a simple example is that due to the connective properties of the neural network (which cannot be replicated mechanically), the more information you add to the brain the more it speeds up, whereas the more information you add to a computer memory the more it slows down. The brain is the most complex known object in the universe – by far.

When Deutsch turns his logic to the subject of what he calls the 'common sense' view of time, he dispatches it remorselessly, first pointing out that

> The idea of the flow of time really presupposes the existence of a second sort of time, outside the common-sense sequence-of-moments time. If "now" really moved from one of the moments to the other, it would have to be with respect to this *exterior* time. But taking that seriously leads to an infinite regress, for we should then have to imagine the exterior time itself as a succession of moments… and so on.

He points out that 'we are accustomed to time being a framework exterior to any physical entity we may be considering', in effect a reference to Newtonian physics where time and space both form absolute frameworks. (Einstein combined them into one framework of space-time). According to Deutsch 'the sequence of moments…does not exist within the framework of time – it *is* the framework of time.' But it is difficult to imagine the sequence being static, as it would be a series of snapshots, where 'the focus of our attention must move along the sequence.'

Pausing here, this concept of the series of snapshot 'nows' is discussed more fully by Barbour who goes some steps further than Deutsch but not inconsistently. The neuropsychology of attention is relevant, because there seems to be a 40 Hz cycle of flashes of attention in the brain, and one could argue that it is 'synchronised' with the series of nows. But one could also argue one step further that it is identical: the snapshots exist *in* our attention, not externally. So our brains impose sequence on the snapshots. Or, if our brains are themselves seen as structure, then the sequence in the brain is part of the same stucture as the sequence of snapshots – part of the same configuration (as Barbour might put it). The sequence is then a pattern in the structure. Sequence is *apparent*, not necessarily real.

Deutsch points out a duality in the ways we think of sequences. 'When we are *describing* events, saying when things happen, we think in terms of a sequence of unchanging moments; when we are *explaining* events as causes and effects of each other, we think in terms of the moving present... No flow of time is involved when we say *when* something happened, any more than a 'flow of distance' is involved if we say *where* it happened. But as soon as we say *why* something happened, we invoke the flow of time.'

Deutsch concludes flatly: 'Physical reality is not a spacetime. It is a much bigger and more diverse entity, the *multiverse*.'

The Shape Universe: Julian Barbour

Julian Barbour's first big book, *The Discovery of Dynamics*, is a monumental elegy to its subject. He starts with Aristotle (having unfortunately only tipped his hat to the 'pre-Socratics', Pythagoras, Heraclitus and Parmenides) and proceeds through to Descartes and Newton. The book is an elegy because with sympathetic (one might say loving) detail Barbour discusses the struggling development of classical dynamics while making it clear that it is now dead. That was clear from the book's original title, 'Absolute or Relative Motion', and its sub-title which contains the words 'from a Machian point of view.' 'Mach's Principle' was invoked by Einstein, but apparently not accurately, and Mach did not set it out as such. Barbour states the 'Machian ideal' as *'there are observable causes of observable effects and everything fits together into a coherent whole.'* This derives from statements by Mach that physics should not concern itself with anything that was not observable (which rules out abstract floating concepts such as space and time) and that the motion of any object was related to the motion of all other objects in the universe. This is what I would call a minimalist position, and Barbour is nothing if not a minimalist. *The Discovery of Dynamics* is a long stripping down of unnecessary concepts.

Near the end (after 700 pages or so which serve as the most complete history of *thought* in physics I know of) Barbour returns to a quote from Mach: 'The object of natural science is the connection of phenomena; but the theories are like dry leaves which fall away when they have long ceased to be the lungs of the tree of science.' As Barbour points out, by 'theory' Mach meant something more like 'explanation.' Barbour is impelled by a similar impulse to let explanations fall away once they have fulfilled their purpose. In a communication to the *Edge* website (2002), he states: 'I have been advocating for a while a dynamics of pure shape. The idea is that the instantaneous shape of the universe and the sense in which it is changing should be enough to specify a dynamical history of the universe.' There is no need in such a dynamics for time, motion, or even space. Mach wrote,

> It is utterly beyond our power to measure the changes of things by time… time is an abstraction at which we arrive by means of the changes of things: made because we are not restricted to any one definite measure, all being interconnected.

Barbour is interested, like Mach, in the connections between things, not distances. And as the words 'instantaneous shape' imply, he is not talking about change in the sense of change over time, he means change *in* the shape. His version of the universe comes close to the geometrical universe proposed by Descartes (although he explains in the Discovery how Descartes sabotaged his own theory for theological reasons and could not develop it properly.) I shall call it simply the Shape Universe, as distinct from the Time Universe. There is room in the Shape Universe for the Multiverse. (Barbour discusses it in terms of John Bell's adapation of Everett's theory into a 'many worlds' theory, and includes it in a 'many instants' theory.). But there would be no room in the Time Universe for the Multiverse.

On page one of *The End of Time* Barbour states that 'Two views of the world clashed at the dawn of thought… Heraclitus argues for perpetual change, but Parmenides maintained there was neither time nor motion. Over the ages, few thinkers have taken Parmenides seriously, but I shall argue that the Heraclitean flux… may be nothing but a well-founded illusion.' Barbour does not return to Parmenides (except once or twice in connection with others, such as Zeno) but instead takes up the Platonic view of 'Forms' (derived almost certainly from Parmenides) and names the Shape Universe he describes 'Platonia.' The Platonic Forms have always been somewhat ghostly entities compared to the world of illusion that covers them. Plato described them as shadows cast on the wall of a gloomy cave by people we do not see passing outside. And Barbour's descriptions of Platonia are of a universe we shall never see.

Barbour starts by describing the simplest possible universe, one of three particles in a triangle, and shows how a more complex universe can be built up by different arrangements of the triangles. 'Each triangle is a possible Now.' And 'whatever the number of particles, they form some structure, a *configuration*...Each Now is a structure.' A larger space which contains many such Nows is in physics a *configuration space*. Barbour uses the image of the configuration space as a landscape crossed by paths of possible histories. 'There is no time in this picture.' He has already discussed classical dynamics and how Copernicus, Kepler and Galileo 'taught us to see motion where none appears. [He gives the example of how Jupiter's retrograde motion is an illusion due to the movement of the earth.] The notion of time capsules may help us to reverse that process – to see perfect stillness as the reality behind the turbulence we experience.'

As Barbour defines it a time capsule is 'any fixed pattern that creates or encodes the appearance of motion, change or history.' An example is a painting, such as those by Turner, in which the illusion of motion is contained. A time capsule is anything that contains a record. It can be our brain, or even the Earth. 'The phenomenon of time capsules is very widespread in the physical world, and is not restricted to our mental states and experiences.' Barbour does not want to, he states, give consciousness any special 'role in the world', and unlike Penrose in The Emperor's New Mind he does not suggest a 'new physics' associated with mental states. Nor does he fall into the trap of David Bohm of involving a supposed quantum mechanics of the brain in his theory. He simply states that 'Consciousness is the ultimate mystery.'

But the result of this parking of consciousness to one side is strangely disturbing. Barbour's minimalism, which I admire, leads to a position rather like Deutsch's – that mind is just one among many things in the universe, in Barbour's theory among many time capsules. There is no difference, in the scheme, between the time capsule that is me and my mind or brain, and the time capsule that is a fossil or a Turner painting. Platonia is above all impersonal. And 'perfect stillness' is in itself deathlike. The past (even our memory of it) is 'never anything more than we can infer from present records.' "The mature brain is a time capsule. History resides in its structure.' Barbour discusses geology and how 'all this fantastical abundance of evidence for time and history is coded in static configurational form, in structures that *persist* [my italics]... The evidence for time is literally written in the rocks.'

Here Barbour's excitement comes across and I am happy to see a poem, for example, as a structure that persists – and to think that in some sense my brain and myself persist and do not die. But Barbour provides a brilliant description of how whenever his family cat Lucy jumps into the air and lands back again, every one of the billions of atoms in her body changes place. 'It is because we abstract and "detach" one Lucy from her Nows that we think a cat leapt. Cats don't leap in Platonia. They just are.' Actually they (and we) are *fields*. 'At the deeper, subatomic level the atoms themselves are in a perpetual state of flux. We think things persist in time because structures persist, and we mistake the structure for substance. But looking for enduring substance is like looking for time. It slips through your fingers. One cannot step into the same river twice.'

Barbour inevitably has a huge problem of language in describing the Shape Universe. Again and again he must use words as they are normally used. So as part of his rejection of the Heraclitan flux he must insist on flux. At one point he writes that 'Time capsules have a cause, but time is no part of it.' Yet in a timeless universe there can be no cause. The Oxford philosopher A.J.Ayer was once asked how he would define cause and effect and replied, 'One damn thing after another.' In the Shape Universe, the damn things are part of a static configuration.

Or else certain sequences of events *in the structure* which we think of as causal are characterised by a particular shape. But we don't, in normal language, describe shape, we describe time. It is a deeply ingrained habit. I.e. somewhere in the structure of the time capsule that is my brain there are sub-structures which we describe as 40 Hz oscillations. Even our sense of time is just part of our structure as we stand on a path in Platonia and see one direction as the past and another as the future. (This is Bolzmann's famous 'clothesline' analogy, which Barbour has explained). *Everything* – even our language – is a structure in Platonia. Deutsch suggests, interestingly, that every language is in itself a theory of the universe. If so, Platonia is cluttered with inaccurate theories. To describe the Shape Universe we shall have to remake our language. Or, to phrase this in conformity with a timeless universe: I don't know if around the next bend of the path the changing shape of language is beginning to correspond with the shape of the universe better than it does now and if part of me is there experiencing it.

On the positive side, the difficulties with language that arise in *The End of Time*, are poetic paradoxes. Barbour, in fact a highly linguistically aware man who knows his Shakespeare better than most English professors and who has been known to spend every evening for a winter reading through German poetry with his wife, cannot help writing well even in the face of the huge task he has taken on. One paradox is that nothing happens in a timeless universe, but that as Barbour puts it we must invent time so that everything does not happen at once. That everything does happen at once is obvious in his vision. He writes 'Creation is here and now.' And in a paper 'On the Origin of Structure in the Universe' (1992), he proposes a 'relational dynamics', based partly on Leibniz as well as on Mach, where the universe consists solely of simultaneous connections among entities ('monads') – all of which are connected to some of each other directly, thus all to all indirectly.

Long before I came across Barbour's work I wrote a love poem which started 'Untime me and untie me, wrench me free.'... and ended 20 lines later with 'Give time to me, / I'll give you mine, entime you, entime me.' In other words no sooner had I got rid of time than I wanted it back. Barbour is made of sterner stuff. He states that 'after seven years of thought' he wrote on the kitchen blackboard: 'The history of the universe is a continuous curve in its relative configuration space.' (Oddly, Robert Graves has discussed in an essay how the Graeco-Roman poet of ideas, Lucretius, describes the universe as infinite because of a constant swerve – 'clinamen.' Graves claims that most modern physics has its origin in Epicurean philosophy, although physicists don't usually know this.)

So it is the structures that persist, not the distribution of matter in each successive Now. The landscape of Platonia is bleak, no matter how well Barbour writes, because it is hard to express geometry (or geometry-like thinking, or a description of a painting for that matter) in words.

> Imagine that Platonia is covered by a mist. Its intensity does not vary in time – it is static – but it does vary from position to position. Its intensity at each given point is a measure of how many configurations (as in the previous example, with triangles in the Timeless Bag) corresponding to that point are present. All these configurations, present in different quantities, you should imagine as being collected together in a 'heap' or 'bag'.... A timeless law, complete in itself, determines where the mist collects. The law is a kind of competition for the mist between the Nows. Those that 'resonate' well with each other get more mist...

This passage is dense with ideas – among others a hint at what Sheldrake calls 'morphic resonance' (since the Nows are shapes) although Barbour does not mention Sheldrake. The word 'competition', although it is necessary to use some such word for the process of selection among Nows, conjures up a world of action and movement. The problems Barbour faces in describing Platonia are huge and the mixing of disparate images makes the landscape seem even more impersonal.

Eventually Barbour digs himself well out of this hole by using diagrams – of corrugated surfaces, of ridges, peaks of wave packets, or mist distribution. And best of all a drawing of how the parachute of a salsify seed where the ribs flare out from a point (another analogy would be a morning glory flower) can represent 'the frontiers of Platonia'. The point of origin of the ribs is point Alpha where the universe looks as if it begins: the Big Bang, in Platonia, is just the point in the shape which is the least complex.

Barbour does have some sympathy with Deutsch's 'multiverse' view. Even Hawking has recently embraced it in *The Grand Design*. But Barbour's is *not* a multiverse theory. He does describe a universe of seemingly infinite (though finite) 'Nows', and states (most recently in a Canadian radio interview called 'Oxford Time' in which Deutsch and Penrose also feature) that each of us inhabits myriads of Nows, so that all of us eternally exist in our Nows. (He remarks that Julius Caesar is as alive as we are). But these are *actual* Nows. He is not describing a multiverse of potential Nows. There is not a Now in which Seán Haldane decides to go to Cambridge, not Oxford. There are only the many Nows of Seán Haldane as he exists in *this* shape universe.

I find I have dwelled on the more complex aspects of *The End of Time*. It is in fact a very clear and readable (and re-readable) compendium of theoretical physics at the turn of this century. It contains a very clear exposition of how 'time is measured by length – the distance traversed by the hand of a clock – and masses are deduced from accelerations, which involve lengths and times.' It discusses views of astronomy from Newton until now; quantum physics in all its major aspects; probability theory in quantum mechanics (probability density being represented as a 'blue mist' over Platonia); and 'intrinsic difference and best matching' – concepts Barbour uses extensively in papers with colleagues (most often Niall O Murchadha) which explore the mathematics that are necessary to explain how the Shape Universe can include or explain the various accepted theories which are used to support the Time Universe and which currently put the Shape Universe into question – the main problem being the evidence from Hubble's red shift that the universe is expanding, which of course a Shape Universe cannot. Barbour and his colleagues are making progress. But the articles (which can be accessed on an Internet search by typing in Barbour and Murchadha) are hard going for those who (like me) are not proficient in mathematics. Barbour's spatialisation of the universe is also inevitably a mathematicisation – an abstraction.

Theories of everything

There is a problem with all cosmological theories. They are untestable. And by definition they are theories of everything. Hawking or Deutsch or Barbour might modestly deny this by saying no theory can encompass *all aspects* of the universe. But that kind of theory of everything would have to be as big as the universe itself – it would have to *be* the universe. The map does not have to be as big as the territory. The scope of these cosmological theories is so large they can explain anything away. It goes all the way back to Heraclitus and Parmenides (as usually understood). *Everything* is flux and change. Or *everything* is a static plenum. Actually, as I think a reading of *Peri Physeos* makes clear, Parmenides avoided the *everything* trap by presenting both sides, as a poetic paradox. It was Plato who froze Parmenides into Forms *only*. So I suppose Platonia is more suitable than 'Parmenidea' would be for Barbour's Shape Universe.

Even in a timeless universe there are, empirically, distinctions and boundaries (or boundary conditions between sets of probabilities). Barbour takes care of this by describing the universe as a 'set of sets'. The multiverse (which he does not discuss) can I think also be accommodated as different configurations in Platonia. But neither his model nor the Time model allows any kind of dualism. The two universes cannot meet. There cannot be a dual universe. Yet here we are, able to experience both universes.

Thomas Huxley in 1865 saw species as equidistant from a common source but not progressing in time – a geometrical view of evolution – rather like Barbour's salsifer parachute universe from point Alpha. Darwin (and eventually Huxley) recognised that this was a static view that led nowhere. The Shape Universe could be called nowhere – because everywhere. As it is Barbour comes perilously close to sounding like Berkeley when he discusses the evanescence of matter.

A flaw of Barbour's theory is that he invokes the 'mystery' of consciousness. In a 'theory of everything' this is not good enough: there should be no areas cordoned off for 'mysteries'. [Jacques Monod explained this convincingly 50 years ago in *Chance and Necessity*]. What generates information? What generates our lives? Is the only answer 'God'? Leibniz and Berkeley took God for granted. Leibniz resorted to Aristotelian 'entelechy' (OED: **1** 'In Aristotle's use: the condition in which a potentiality has become an actuality; *spec.* the essential nature or informing principle of a living thing; the soul.... **b** *Biol.* A supposed vital principle that guides the development and functioning of an organism... **2** In Leibniz's use: a monad.) 'Entelechy' was taken up by the vitalistic biologist Driesch and informed Reich's 'orgone energy'. The word Platonia opens the doors to metaphysics on the largest of scales. A scientific theory of everything is surely at risk if it implies God. Or do all theories of everything imply God? Wilhelm Reich cracked grandiosely that he had put God in the laboratory, but in prison he prayed humbly to the 'cosmic energy ocean'. 'Soapy Sam', Bishop Wilberforce, explained patronisingly to the Darwinians that evolution was all in God's mind. Barbour remarks that the world of our experience (as distinct from the underlying static universe of forms, Platonia) is an illusion – all of it, not only the illusion of time. This is not far from the idealism of Berkeley's God dreaming the world. Lewis Carroll, a mathematician who anticipated the theory of relativity in his humorous writings, satirised this in Alice in Wonderland. Tweedledee says to Alice about the Red King, sleeping and snoring beneath a tree: 'Why, you're only a sort of thing in his dream!' 'If that there King was to wake', adds Tweedledum, 'you'd go out – bang! Just like a candle!'.

I think Barbour has a religious attitude to the world, and I suspect that at times he is unhappy about the godless nature of Platonia, but I do not think he wants to put God there. (If he did he would say so). He rejects a plenum: the Shape Universe as he describes it is a huge pattern of sub-patterns, a unity of unities, a set of sets – one of which is each of us. But his idea of 'universal simultaneity' in a universe which is not expanding is surely close to a plenum. He may be closer to Parmenides than he thinks.

In the last chapter of *The End of Time* Barbour allows himself to think more speculatively and reveals some of his human ambivalence towards his intellectual vision. 'Is Platonia a graveyard? Of a kind it undoubtedly is, but it is a heavenly vault.' Barbour can think poetically, and can hold two sides of a paradox in his mind at once. Again in his more self-indulgent last chapter he comes back to Heraclitus and Parmenides, remarking that they represent 'verbs' and 'nouns' respectively, and that it is impossible to think without both verbs and nouns. (This is similar to an idea of Goethe's that 'Being is Doing') Barbour writes:

> If my definition of an instant of time is accepted, it becomes hard to say in what respect those two great pre-Socratics might differ. The two best known sayings attributed to Heraclitus are "Everything flows" (*Panta rei*) and the very sentence which, entirely unconsciously, I used to clinch the argument that the cat Lucy who leapt to catch the swift was not the cat who landed with her prey: "One cannot step into

the same river twice." There is always change fron one instant to another – no two are alike. But that is just what I have tried to capture with the notion of Platonia as the collection of all distinct instants.

But who or what is the 'one' who steps into the river? An instant stepping into an instant? Just as Parmenides could not help, it seems, including an urgent vision of life and movement in his account of the unchanging plenum, Barbour cannot help allowing life into the 'graveyard' of Platonia.

Free Will

All three universes described above are deterministic to some extent. And all three physicists discussed above arrive at the question of free will at the end of their books.

The Time universe is the most deterministic of the three. It is like a giant clock ticking away, powered by a spring that is gradually weakening. It began with a point of potential energy and this energy is running down as the universe expands further and further. Every moment in the expansion from the Big Bang is determined by the moment before. Of course it is so unimaginably complex that we cannot predict what will happen in the next moment of existence. So we feel as if we have at least some choice in determining it, but this is an illusion. At any point in the universe time is like a river, sweeping everything along with it. The only way of escaping it is through actions that seem to prolong or suspend it, like meditation or listening to music, but these illusions are themselves determined by what leads up to them. And you and I are determined by a genetic code which replicates itself by determined laws and by the action on us of the world around us, determined by its previous history. We may debate about how much we are determined by heredity and how much by environment but that is just part of the complexity of the universe. That we are determined is not in doubt. Hawking, even when weighing up the evidence in quantum physics for uncertainty in determining what is happening, concludes: 'It is just that we try to fit the waves to our preconceived ideas of positions and velocities. The resulting mismatch is the cause of the *apparent* unpredictability.' [My italics]. In addressing Einstein's question, 'How much choice did God have in constructing the universe?' Hawking states that 'If the no boundary proposal is correct, he had no freedom at all to choose initial conditions. He would, of course, still have had the freedom to choose the laws that the universe obeyed. This, however, may not really have been all that much of a choice.' Here Hawking is being mischievous again. He does not believe in God except as a theoretical answer to the question of *why* the universe began.

The Multiverse contains a sort of parliamentary version of free will in which the extent of it is determined by a majority among millions of one's selves in different universes. Deutsch side-steps determinism neatly:

> The difficulty of reconciling free will with physics is often attributed to determinism, but it is not determinism that is at fault. It is… classical spacetime… Spacetime does not change, therefore one cannot, within spacetime physics, conceive of causes, effects, the openness of the future or free will…. Replacing deterministic laws of motion by indeterministic (random) ones would do nothing to solve the problem of free will, so long as the laws remained classical. Freedom has nothing to do with randomness… Who would value being random?

Deutsch gives examples of free will in the multiverse. 'After careful thought I chose to do X' becomes, 'After careful thought some copies of me, including the one speaking, chose to do X.' And 'I am good at making decisions' becomes 'The copies of me who chose X, and who chose rightly in other such situations, greatly outnumber those who did not.'
One has to admire Deutsch's verve in pursuing the implications of quantum physics so far. As mentioned earlier, his very literal interpretation of the quantum interference experiments is not widely shared – perhaps because the implications are so bizarre. But his logic is impeccable. It is only that, as in his discussions of computation and mind, his conclusions do not seem to include life. Free will in the multiverse does not *feel* free. It feels passive. David Deutsch may cheerfully write a book in one universe and make a

cup of tea in another parallel universe, and so on for ever – not a time 'ever' but a sort of lateral 'ever', as in ever repeating waves on an ocean, all existing at once. But if there are billions of each of us, how is it that we feel we are *here?* The answer that there are billions of us feeling we are in billions of heres is unsettling to say the least.

The Shape Universe is also determined. It must be: it does not contain time, cause-and-effect, energy, or movement. It just *is*. How can we possibly exert choice where everything is happening at once? All of our apparent choices already exist somewhere in the configurations of the Shape Universe, but since they are in time capsules we have the illusion of sequence and of making them. This *illusion* of making choices occurs in the Time Universe too.

Barbour states:

> From personal introspection I do not believe that my conscious self exercises free will. Certainly I ponder difficult decisions at length, but the decision itself invariably comes into consciousness from a different, unconscious realm. Brain research confirms that what we think are spontaneous decisions, acts of free will, are prepared in the unconscious mind before we become aware of them

This is only partly true. For one thing in pondering a decision a sort of mental trial-and-error (or as Popper puts it 'conjecture and refutation') occurs, in which predispositions or impulses from the unconscious (perhaps mainly sub-cortical structures) have it out with critical thinking from the conscious (perhaps mainly frontal cortical structures) in a process of dynamic interaction. For another thing, the dispositions and impulses which are mediated in sub-cortical areas (in pathways from the peri-aqueductal grey areas around the brain stem to the limbic system via the regulatory centres of the hypothalamus) are themselves influenced by autonomic 'conditioning' and by stored memories of experiences which may themselves have been conflictual. And the whole dynamic system varies in its activation according to emotional arousal, so that for example the more excited and aroused the person is, the more the sub-cortical information will over-ride the cortical – and vice versa. But a strong point in Barbour's favour is that muscle potential for a specific action is activated before the person is even aware of making a decision.

Decision-making in brain terms cannot be described except with reference to structures, and the Shape Universe theory, being an all or nothing theory, can provide an explanation for *anything* in its own terms. In this case, apparent conflict between brain structures could probably be set out in a series of diagrams, i.e. it can be spatialised. In a universe without time, the conflict is just a particular kind of shape.

Barbour also leans on his version of the many worlds (multiverse) theory.

> However the many-instants interpretation puts an intriguingly different slant on causality... In both classical physics and Everett's original scheme, what happens now is the consequence of the past. But with many instants, each Now 'competes' with all other Nows in a timeless beauty contest to win the highest probability. The ability of each Now to "resonate" with the other Nows is what counts. Its chance to exist is determined by what it is in itself. The structure of things is the determining power in a timeless world....
> I do not think that we are robots or that anything happens by chance. That view arises because we do not have a large enough perspective on things. We are the answers to the question of what can be maximally sensitive to the totality of what is possible.

Again causal language gets in the way, although Barbour neutralises it with quotation marks, in how the Now 'competes'. But the paradox of the Shape Universe is that this static plenum is more an 'existential' universe than the Time Universe.

In the Time Universe our present is a prison determined by the past. We don't know what will happen next but our life must follow the 'laws' Our only freedom is the false freedom of not knowing what our jailers – the programmes of our genes or the determinations of past events – have in store for us next. Meanwhile, politicians, like the proverbial 'jail-house lawyers' rant about 'the laws of history', the inevitability of events, and 'destiny.' Character (the individual) only has an illusory destiny.

In the Shape Universe everything is happening at once and in a sense already has happened: past, present and future co-exist. Even evolution is not an unfolding over time, but a succession of shapes in space. (Curiously, the genome as described by Matt Ridley, functions according to the *placement* of genes. Seen without time, the genes do not determine our development they coexist with it).

Past, present and future co-exist everywhere and most interestingly they co-exist in mind. We do not know what will happen next but we have intuitions, hopes, desires etc. which *since there is no time* are *making our lives happen*. Where everything is simultaneous and interconnected we each contribute to the whole. We are keeping the whole thing together. And we do so consciously. Leibniz's monads are conscious. And in the Machian universe nothing is real except what is *observable*. We are observers and observed, and this interaction gives us a dynamic existence. Mach said 'Human existence has nothing to do with "things", it has to do with *awareness*.' ['Bewusstseinhalten'] As Heraclitus is supposed to have first said (although it is at odds with his universe of flux), 'Character is destiny.' The Old English word for 'fate' was 'wyrd' – 'becoming'. At each instant of the Shape Universe we become ourselves. And, as Barbour proposed in a paper on Leibniz, 'Because of the way in which experiences are generated, we are all continually sharing experiences, although there is never identity of experience. In fact the entire world is resolved into *pure shared experience.*'

Barbour *realizes* Platonia – makes it real. He does not appear to be an idealist. He would probably, as an astronomer since childhood, agree with David Stove that 'If you lie face-up in the open air on a clear night, you are suddenly reminded that, in a line drawn from your face outwards, there is nothing, however near or far, which takes or even could take the smallest interest in you.' In fact Barbour, the 'gentleman physicist' as he has been called in reference to the facts that he lives in the country and does not work for a university, has eliminated Platonic idealism from 'Platonia.' 'Is it a graveyard?" he asks.

Barbour's Platonia is, I think, misnamed. It is not Platonic. It is a motionless, spatialised and geometric Leibnizian or Machian universe in which everything exists in relationship with everything else – a 'bootstrapped universe' which Leibnizian 'monads' cooperate to hold together. In a conversation with Barbour I rather naively expounded Karl Popper's idea that reality is an inter-subjective consensus arrived at by observers in the absence of evidence to refute their hypothesis. I used the example of several people stationed around the edges of a field observing a cow from different angles and comparing notes, combining their perspectives. Barbour remarked: 'But the reality has to include the cow's perception of *you.*' So the universe becomes the sum of all possible perceptions.

Finally, in the End of Time, Barbour lets the cat (or the 'field' called the cat) out of the bag and states about the many worlds theory: 'However, despite strong intellectual acceptance of many worlds, I live my life as if it were unique.' In other words, he admits to making a *choice.*

The French philosopher Emile Boutroux made the point that free will must exist because we are capable of thinking it does. And philosophically the question of *how much* free will we have is unimportant. It may be impossible to calculate our 'degrees of freedom' but the 5% or 95% we have are where we are free. The Multiverse and the Shape Universe at least include awareness. As Colin McGinn has said, any philosophy that denies the obvious must be suspect. What could be more obvious than time? Of course it *exists* – but not as an absolute framework or dimension in two of the three universes of modern physics. In the Shape Universe it is structural. Paradoxically, we become aware of this structure in persisting *against* it – and we ourselves are structures. As we persist, each in our own unique way, we create the universe. Give me the Shape Universe any day, over the others! Just don't kill it by calling it Platonia.

3

MORE-THAN-COINCIDENCE

Miss Love
I was born into a rational tradition in which, however, the irrational or unexplained was accepted without any fuss. I was baptized into the Church of England but my mother, who was only 19 when I was born, had simply never believed in God. Nor had her mother who was descended from German 'free-thinkers', nor her English father who never went even to the village church. My grandmother had, however, several times seen the ghost who inhabited the ancient house we lived in: dressed in a navy blue eighteenth century jacket he would occasionally emerge from the wall in her bedroom then walk out of the door into the hall and disappear. This never bothered her. When my Irish father returned from the war we moved to Northern Ireland. He belonged to the Church of Ireland, more austere than the Church of England. I went to church to sing in the choir and join the Wolf Cubs, but did not want to be confirmed. My father did not mind at all: he saw confirmation as an adult and reasoned act. He believed in God and in an afterlife, but he thought for himself – true to the best in the Church of Ireland which, as the essayist Hubert Butler pointed out, has a tradition of emphasising 'the right to private judgment' and even 'the right to be wrong'.

My father's most frequent admonition to me was 'Think, Boy, think.' As I grew up he gave me the run of his bookshelves where I read Joad's *History of Philosophy*, Freud's *Totem and Taboo*, and J W Dunne's *An Experiment with Time* which set me to writing down my dreams. I don't think I did this long enough to be certain if my dreams were predicting the future as Dunne said they would. I remember my experimental attitude more than the results. At age seventeen I wrote a long poem to my girl-friend which startled me in that it seemed written by someone else using my hand, as if automatic writing – although it was undoubtedly about me and as I wrote it I felt shivery and tearful. I had been in the science 'stream' at school but found it boring and had switched to English and languages. I had no explanation for 'inspiration', although eventually I found it was taken seriously in essays by Robert Graves.

At Easter 1961, when I was aged 18, I visited my grandparents in Sussex, then a school friend who had moved to Wales. When passing through London I bought Graves's *The White Goddess*, just re-published after a long period of being out of print. On the last night at my friend's house when I turned out the light in my bedroom, in the pitch dark a huge glow of white seemed to advance on me from the window. It was a blossoming fruit tree outside. The word 'mulberry' came into my head. This seemed unlikely. In curiosity I turned on the light and looked in *The White Goddess* which contained chapters about tree lore. I read there, not surprisingly in view of Graves's theme, that mulberry was 'a fruit sacred to the White Goddess'. The next day, returning to Ireland via Liverpool, I found myself sitting on the train opposite a remarkably pretty girl a few years older than myself. She had unusual yellowish brown eyes and honey coloured hair, and was looking out of the window. I wished I could find some excuse to start up a chat with her. In the hedges rushing past the train window were white blossoming trees. Reminded of the tree of the previous night and looking for one similar, I wished, idly, that the girl would produce a book on trees or something, so that I could make conversation about blossoms. She leaned down and pulled a book from a small bag at her feet. A botany book. Seizing my chance I asked her whether her book had anything about trees, because I had recently seen a tree that might have been a mulberry in blossom. I described it. She looked in her book and said she thought the tree I had seen must have been a pear. She was polite but cool, with obviously no time for a mere boy. She started to read. I found myself wondering what her name was. A sort of love-goddess, I thought romantically. Then I decided, for no reason, that her name must be Fiona. This was the name of a close cousin with whom I had lived as a child in Sussex, but this girl did not resemble my cousin at all. When we arrived at Crewe junction I took the opportunity to help her by lifting her suitcase down from the luggage rack. I then saw a label on the side of the suitcase which had been turned inward: F.Love, Trinity College, Dublin. I was so flustered that I lost sight of her in the station. In the train to Liverpool it was crowded and I had to stand in the corridor, near a young man with black hair, a 'lounge lizard' I thought, to whom I took an instant and inexplicable dislike, so much that I had to move away. At Liverpool station I saw him again, running toward the young woman. They greeted each other with a hug and set off toward the Dublin boat. I was going to Belfast where I arrived the next day in a state of excitement which resolved itself in a rather naïve poem:

> Who sent you to me
> Death-white beauty of blossom?
> Why was the night illumined by your glow?
>
> Who sent you secret stranger
> Whose pale smile
> Sent icicles of pleasure to my heart?
>
> What rival lured you back into the night?

The next day, haunted by this incident, I looked up the girl's name in the Trinity College register. It was Fiona Love. As I reached the end of the White Goddess I read: 'Chains of more-than-coincidence occur so often in my life that, if I am forbidden to call them supernatural hauntings, let me call them a habit. Not that I like the word "supernatural"; I find these happenings natural enough, though superlatively unscientific.'

I felt I had encountered the White Goddess herself on the train. But when subsequently I experienced similar chains of more-than-coincidence (which I did, and do, to the extent that I can agree with Graves that this is a habit) they were not especially associated with emotionally charged events. They do sometimes occur immediately preceding the writing of a poem (which is not, for me a voluntary act, and can sometimes feel like the final link in the chain). But they can also occur any time and are often banal. Even the more-than-coincidence of the Miss Love episode turned out to have a link which had nothing to do with poetry: my cousin Fiona (to whom I said nothing about Miss Love) some years later married a graduate of Trinity College Dublin who had been there at the same time as Miss Love.

Coincidence and more-than-coincidence

I met Robert Graves later that year in Oxford where he was visiting professor of poetry. He remarked, in a discussion of how poems were sometimes written years away from what they described, 'There is no time, you know.' He had written somewhere of 'time's ineluctable wibble-wobble' and like so many people alive in the 1920s and 1930s (including my father) had been exposed to J W Dunne's ideas about time being 'relative' and 'seriality' – which echoes *The Law of Series* set out by the Viennese biologist Paul Kammerer in 1919. It is not known if Dunne had read Kammerer – any more than it is known for certain that Karl Jung had. But Jung's much better known concept of 'synchronicity' is without doubt (as Koestler demonstrates in *The Midwife Toad*) based on Kammerer's work.

More crucially, in the 1920s and 1930s Graves had lived and worked with the American poet Laura Riding who wrote a Leibnizian poem about the 'slippery monads' and declared that time had stopped: 'All the politicians who are going to be elected have been elected; and all the artificial excitement in events which no one really regards as either very important or very interesting has been exhausted. All the historical events have happened.' All that remained for thinking people was to make true 'judgments' about what had already happened. This in 1935. (Step forward – in 1992 – Francis Fukuyama, with *The End of History*.) Questions about time were in the air between the World Wars. Wyndham Lewis's best-seller *Time and Western Man* (1927) attacked the 'time cult' of remaining up to date and living the present moment for sensation only, not pausing (against time, as it were) for thought. Time seemed to accelerate at the end of the 1930s: World War II loomed, and ironically Laura Riding ran off from Graves to live with a correspondent to *Time* magazine. As the various time fads of between the wars subsided Graves clung to his pragmatic 'more-than-coincidence'. It fitted the facts of his existence. Many people experience it. Unfortunately more-than-coincidence and 'seriality' (or synchronicity as many call it, following Jung) are easily dismissed as illusions due to the statistical laws of chance. Sooner or later – since so many billions of things happen in the world – events that appear to be unusually connected will occur 'by chance.' But this is too big a statistical game to be scientific. The laws of chance even in such limited areas of action as roulette or card games are not – probably cannot be – understood, since they operate on a universal scale. Thomas Hardy wasted time in the mid 19th century studying the book of the day, Moiré's *Laws of Chance*, but abandoned this in novels and poems which simply reflected his own experience. He was condemned for relying on 'coincidences' too much in his plotting. The same criticism has been made of Anthony Powell's 12 novel cycle *A Dance to the Music of Time*. But the recurrence and co-occurence of themes, people and events are true to life. An examination of almost anyone's life suggests that coincidences and events where history appears to come around on itself are not rare. I have seen this frequently in taking clinical histories. Coincidences (though not always 'more than') happen to us all.

I had two main girlfriends in my teens, in Belfast, in the 1960s. When living in Victoria, BC, in the 1980s I found out that one girlfriend's favourite brother was a near neighbour. And I met two other people, separately, who knew the other girlfriend. It is claimed that everyone in the world knows everyone else by five removes. And plenty of Irish people emigrate to Canada. Even more unusual coincidences may have a surprise value but do not necessarily defy the usual laws of nature. For example, in Victoria I met a 70 year old Irishman from Dublin and got talking about the West of Ireland. It turned out that when I had spent summer holidays at Dog's Bay in Galway with my parents who hired a caravan and drove it there, with me camping in a field in a tent, this man had also spent holidays there with his family, parking his caravan in the next field. I mentioned one thing that had haunted me. There was a beautiful girl of about 14, with blue eyes and black hair, who worked in the butcher's shop in Roundstone and was always clothed in black, obviously in mourning. I had never dared talk to her. I asked the man if he had ever seen her. He said he had known her and her family and the mourning was because one of her brothers had drowned in Roundstone harbour. This somehow completed the story. But I don't think it counts as more-than-coincidence. Lives interweave each other, that's all.

'What a coincidence!' You are thinking of Josephine and there she is coming around the corner. Or you find someone at a party who shares the same birthday. These may seem 'more-than-coincidence' if they are emotionally charged. You have been fretting all morning about whether to ring Josephine up and ask her for a date. Or you already sense an affinity with the person at the party. A coincidence often consists of a thought followed by an event. Jung saw such coincidences as events in the so called 'collective unconscious'. His term 'synchronicity' means 'at the same time', which is not accurate. You think of Josephine, *then* she appears. You mention your birthday, *then* the other person says 'Me too!'. Paul Kammerer in *Das Gesetz der Serie / The Law of Series* defined such coincidences as 'series'. One event is followed by another and perhaps even by another and another, all of which resemble each other.

The law of series
Kammerer collected examples of series, in his own life and from friends and relatives, for many years, and in his book he set out a classificatory system for them. Classification was second nature to him, as a biologist. Unfortunately, a capacity to overstate and even fabricate evidence also seems to have been second nature. He eventually shot himself in disgrace over 'The Case of the Midwife Toad', as Arthur Koestler's book on Kammerer describes it. He probably did not, in fact, clumsily fake markings on toads' feet, with India ink, in order to support his case for a Lamarckian theory of inheritance of acquired characteristics. It is possible that some one else, perhaps his arch rival William Bateson (Gregory Bateson's father) did so and framed Kammerer. But it was the kind of thing Kammerer himself might have done. Alma Mahler, the composer's widow and femme fatale of early 20th century Vienna (she had affairs with the artists Klimt and Kokoschka, and married the writer Franz Werfel and the Bauhaus architect Walter Gropius) was for a while Kammerer's laboratory assistant and (inevitably) lover. She wrote that she had often found it necessary to constrain him from exaggerating his experimental results. And she would know.

None of this affects the credibility of Kammerer's classification system for 'series', or the validity of the concept. But the suspicion inevitably arises that he may have made some up of his more striking examples. And his reputation was so shattered by his suicide, widely seen as an admission of guilt although he was apparently depressed because his mistress had refused to follow him to the Soviet Union where he had been invited to teach. His best seller *Das Gesetz der Serie* swiftly dropped from sight, and has never been translated into English. So as not to raise questions about the credibility of Kammerer, I shall provide my own examples. If the reader thinks I am making them up, I can only deny this and invite the reader to take notes of series in his or her own life as promptly as possible after they occur.

What counts is the person's observations – even where the theory is wrong. Often a person's theories are so bizarre that their observations are put in question and they are accused of charlatanism – as with Mesmer, Reichenbach, Reich, and Kammerer. Even J W Dunne, who could not have been more serious and less self-interested, was dismissed as a crank. I hope I am not a crank. I can hardly be mentally ill having worked in public mental health systems very successfully for over 30 years. I cannot be all that weird if I manage a clinical service, work day in and day out as a neuropsychologist in a memory clinic with specialist medical colleagues, survive long cross examinations in medical-legal work, and supervise numbers of clinical psychologists at doctoral level. I don't make claim to any huge theoretical advances. I am naturally sceptical and resistant to 'confirmation bias.' I am just trying to explain observations and experiences which are not adequately explained elsewhere and which presumably are difficult to explain – otherwise the once reputable scientists mentioned above would not have been discredited once they offered explanations. And admittedly their explanations are bizarre. But the obstinate presence of what I shall call 'precognition series' and simply 'series' remains.

Precognition Series
When a series involves not simply two or more observable events (same birthdays) but one mental event and one observable event (thought of Josephine, appearance of Josephine), the question arises of whether it is in fact a precognition. I call this sort of series a 'precognition series.' I will start with some examples of the dream precognition that aroused my interest when as a teenager I read Dunne. Dunne is not usually accused of faking his data, even by those who think his books are rubbish. But rather than use his own copious examples of dream precognition I will use my own – again in the hope of encouraging the reader to examine his or her own experiences. I shall add examples of waking precognition.

The function of dreams has unfortunately been obscured in the 20th century by Freud's fantasy that they expressed unconscious or repressed desires in symbolic form: you dream of a snake and it is a penis, etc. And the psychoanalyst is the priest of the unconscious who can interpret the dream. In fact Freud's view is the opposite of the truth – as it is in other areas, such as his claim that memories of childhood sexual experiences and abuse are repressed, whereas as any clinician can attest (or common sense can), that such memories are usually so *un*-repressed as to cause obsessive thinking about them: they are impossible to escape. More pragmatic students of dreams in the 20th century included the poet Robert Graves as well as Dunne, both of whom saw dream experiences in a continuum with waking experiences, the difference being that the former provided uncensored information – again the opposite of Freud's fantasy. This view is supported by the scientific dream research, such as J Allan Hobson's, which finally pushed Freud aside. Hobson and others describe and record the pulsatory alternation during sleep of periods of REM (rapid eye movement) sleep and longer periods of non-REM sleep. During this alternation brain activation switches back and forth with changes of the chemicals which mediate neurotransmission (mainly acetylcholine in REM sleep and serotonin with dopamine in non-REM sleep). Hobson describes dreams as similar to psychosis, hallucinations, and delirium in that conscious thinking is not in control, but nevertheless they contain real information that the brain needs to sift and organise during sleep – whose function it is to permit this. The whole process can be seen as part of the constant orientation (see the final chapter of this book) which is our central experience – or necessity – in a timeless universe. Dreams orient us as do our waking thoughts. Such phenomena as precognition and 'series' can occur both in dreams and waking and demonstrate the continuum between these.

Dreams include openly sexual and bizarre experiences along with a jumble of emotions, some of which are 'surreal' and disturbing precisely because the 'lid' of repression or censorship is off, not on. To experiment with apparent precognition in dreams, where most examples seem to be found, the reader need only record dreams in detail immediately on waking, being sure to take note of *all* imagery, no matter how seemingly irrelevant to the dream's main theme. If you have difficulty remembering dreams, you might follow Dunne's suggestion for retrieval, by casting your mind back to the first thought you had on waking up, rather than to what you might have dreamt. The waking thought leads back to the dream, like a rabbit running back into its hole. It should be enough to note dreams for a few nights only, but every evening run over the whole collection of accounts. You may notice imagery of events which have occurred subsequent to dreams. (Dunne claimed that dream imagery refers about equally to current, past and future events – in thirds as it were). Or the accounts may serve to check a 'deja vu' experience you may have had during the day.

The following accounts are given either without comment or with brief comments only. Discussion will follow. For ease of reference these accounts are labelled P1, P2 etc. – P for precognition, although in some examples precognition is less clear than in others.

P1

In 1976, when living in Quebec 50 miles from Ottawa, I re-read Dunne's *An Experiment with Time* and decided to make an effort to record my dreams. The first night I dreamt of a very dramatic fire in a wooden-floored grocery store where I was buying tuna fish. I noted this on waking, thinking casually that the only grocery store with a wooden floor which I knew had burned, about a year previously, was Devine's in Ottawa. That day I had to go to Ottawa, and following Dunne's precepts, I bought a newspaper and read it carefully for evidence of stores burning down, but there were none. Later in the afternoon I forgot about the subject, and was standing on the sidewalk waiting for a friend when I spotted someone I vaguely knew coming toward me. It was Mr Devine whom I had not run into since before the loss of his store. We exchanged chitchat. He mentioned that a building under construction just behind me was a new business property (though not a store) of his. This meeting of Mr Devine seemed to me to fulfill the precognition of the dream image – though indirectly. I had thought I might encounter a reference to a store burning down, not that I would encounter Mr Devine.

I made note of this incident, and have copied it for the above paragraph. But there is a sequel. In 1978, the new building in front of which Mr Devine and I had met also burned down.

If I dream of Mr Devine's burning store (P1) and the next day run into Mr Devine in front of his new building which one day will burn down like the previous building, it appears that even as I experience the first event of the series in my dream, all subsequent events are already in motion, or already have occurred. I know, though I do not consciously think about it, that I will be in the area of town next day where Mr Devine's original building was and which is his 'territory.' Mr Devine is also involved in the finishing of his new building. It will burn down some day although he does not know this – because it is his destiny, he is 'fated', to be involved with buildings that burn down. He is not an arsonist, but burning buildings seem to be part of a series for him.

The fact of Mr Devine's store burning down is also part of a very long term series for me: in my life I have known no less than three quite bald grocery store owners whose stores have burned down, which I think is unusual statistically.

The first balding man was the proprietor of the Cavehill Road Post Office and general shop, in Belfast, who was suffocated to death in the smoke of his burning shop in, I think, 1949: I watched the fire and saw him being carried out under a blanket. The second was Mr Devine, in Ottawa, in early 1976 as described above. The third was the proprietor of the grocery store in Ladysmith, Quebec, which burned down some while after I moved away from there at the end of 1976. All three buildings had wooden floors.

P2

Another precognitive dream from 1976, the night after the dream in P1:

I dreamt of Singapore and of a huge Zeppelin balloon. The following day I was looking through magazines with my eldest daughter. One had a huge fold out picture of various balloons, including some Zeppelins. Another had an article on guppy-harvesting in Singapore. The guppies were striped like the balloons.

This kind of banal, undramatic series seems to occur frequently with dream precognition. This particular series, however, does seem to have been triggered by the fact that the previous day I had heard the Premier of Singapore, Lee Kuan Yew, talking on the radio. He had made an amusing remark: 'the grass gets trampled when elephants make love.' Perhaps the bloated body of an elephant suggests a zeppelin. But 'triggered' suggests a cause. Perhaps my hearing Lee Kuan Yew was simply the first event in the series.

P3

In September 1983, on Vancouver Island, I go with my wife and daughters to the local agricultural fair. Low on cash, we have to empty our change jars for funds. On the way I remark that just as last year when some acquaintances, T and J, were also at the fair on a Sunday morning, we will probably run into them.

In the goat barn with my smallest daughter I notice another acquaintance of my wife's. I think, 'Oh God, I can't remember her name. What is it – Margaret or Marjorie? At the same instant a farmer who has entered beside me with a goat says loudly 'I'll tie up Maggie here.' Oh yes, the woman's name is Maggie.

Later, while looking at some pigs I run into T with one of his daughters. 'I thought I'd run into you', I remark, 'since we seem to follow the same routine year by year.' He then jokes that last year when he ran into me my wife and eldest daughter were away in the washrooms. This year *his* wife and daughter are in the washrooms. He later remarks to my wife that he and J have been so short of cash they have had to empty their change jars.

I suppose another element of 'series' is the fact that both sets of acquaintances (the only ones we ran into at the fair that year) are couples, like us, with two daughters.

This is mainly an intricately involved set of series, as described in many examples by Kammerer. The precognitive element – my guess that we would run into T and J – is probably only apparent, since we do share the same routines of going to the fair.

P4

D, a woman friend from Vancouver came to visit us in Victoria with her new companion, to introduce him to us. She had not visited since two years previously. At that time she had a friend, Y, in Vancouver, but they had broken up and I had recently heard he had moved to Victoria and was living with another woman I knew. Before D arrived, I found myself thinking idly that if we went out downtown we might run into Y. How embarrassing that would be. On the other hand I had not seen Y since he had moved here. As it happened we did not go downtown with D and her new friend. But in our house my wife suggested I show them my new consulting room. I took them downstairs and we walked into the room. The telephone rang and I picked it up. 'Hello. This is Y. I wanted to make an appointment to see you.' I told him hastily to call me back later because I was busy. Then I showed the room to D and her friend.

This is an event of 'more-than-coincidence' involving D and Y, although neither knew about it. I had had a kind of premonition, difficult to distinguish from precognition.

The subjective experience in series

The differences between precognitive and other series will be discussed later. But the basic distinction is already being made in the word 'precognition' itself. This 'knowing beforehand' is consciously labelled, since the person thinks 'This or that may happen, this image may be a precognition.' If this moment of expectation does not occur, then there is no precognition, only a series of coincidences or striking 'more-than-coincidences.' This definition depends on the perspective. Looking back from event B to event A which seems to have heralded it, event A is merely the first of a series. Looking from event A forward to anticipated event B can be experienced as precognition or, more vaguely, as a premonition. But it is difficult in life to distinguish precognition from fear, premonition from anxiety. Only when event B occurs is the precognition or premonition 'confirmed'. And even then we may ask ourselves if we have made self-fulfilling prophecies.

Emotion seems to play an important role. But at the time of event A it is not necessarily present: indeed a precognition (cognition means thought, not feeling) is unemotional, and can become swamped by fear or excitement. Once we know we are emotional about something that may happen we seem to feel that perhaps we make it happen or draw it upon ourselves.

'I just knew it would happen' may describe a precognition or an anticipatory fear. The most clear examples of precognition are in themselves non-emotional, as in the often neutral imagery of a dream or of a mental flash – although more strictly they can be described as pre-emotional, since the realisation of a precognition when event B occurs, is usually accompanied by emotion.

Sometimes this emotion lends an almost magical quality to the series of events, which seems to follow an emotional, associational logic. Take my example of 'Miss Love' on the train. This was overwhelming to me at the age of 18 and disturbed me so much that a naively unskilled but 'true' poem emerged a few days later in which the imagery comes from the series.

This sort of process is what Graves means by the 'chains of more-than-coincidence' which he has noted often precede a poem. The poem is often a surprisingly literal account of what has happened: its images are juxtaposed by events in emotional life, not by deliberate art. In fact the poem can be seen not only as a record of the 'chain' but as its final link – the last event in the series (see Chapter 7). It might seem that my excitement faced with the strikingly attractive Miss Love in some way enabled the series to occur. In my experience some precognition series involve emotion. One trivial example, in a sense emotionally opposite to the Miss Love episode, was when, visiting Vancouver where I had not been for some months as I went into a large self service restaurant I found myself thinking 'If there's one person I do NOT want to run into today it is X.' Then I joined a queue moving slowly towards the food counter. As the queue curved around towards the counter I could see several people in front of me, and there was X.

Event series

Here are some series in which the first event is not mental but physical, which have not been subjectively considered to be precognitive, and which have been associated with no special intensity of emotion apart from brief excitement or surprise when the series completes itself. I shall call them 'event series' and label them as 'S'.

S1

In a bookstore on Vancouver Island I find myself looking at a beautiful photo of Cornwall which stimulates me to look up a number of travel guides. I have the idea that if we manage to spend some time in England it would be nice to live for a while in Cornwall, a county I have never visited. Penzance in particular sounds attractive, but the books do not give much information about it. The next day my wife receives a call from a friend from back East who is passing through town. My wife invites her to stay the night. They have not met in three years. The woman is originally from Penzance, and is the only Cornish person I have ever met in Canada. So I can ask her all the questions I want. My daughter C wants this woman to read to her from her latest favourite book, a version of Robin Hood I have found in the Library earlier in the week. It is written by Bernard Miles, an actor who specializes in a Cornish accent.

This is a very simple series involving one element: Cornwall.

S2
My four year old daughter was talking one morning about a little girl of eight called Shawna, grand-daughter of our neighbours, who had recently been visiting. My daughter was intrigued by the similarity of the two names Shawna and Sean. Then after lunch she began asking where Shawna was now. I explained she was back home in Alberta with her parents. The telephone rang, and my daughter ran to answer it – her latest achievement. A conversation followed, as if child to child. 'Who is it?' I asked. 'She's called Shawna.' My wife went and persuaded my daughter to relinquish the telephone. 'Who is this?' my wife asked. 'This is Shawna. Could I speak to that little girl again?' My daughter returned to continue the conversation. But this was *not* our neighbours' grand-daughter, who was 700 miles away. This was another Shawna, a child just dialling around at random.

S3
One morning I am thinking of an old friend, James Reeves, and an anecdote about him and the actor David Niven when they were at the same school. At lunchtime I buy a newspaper and read that Niven has died. Later in the afternoon I am reviewing the casenotes of a patient who is called by the unusual name Livia. When I arrive home I turn on the radio. They are doing an extract from a Sheridan play, The Rivals, and within a minute or so a character enters calling out: 'Livia!'

This begins as a precognition series, which is then linked to a second series – in this case by the recurrence of the same rather unusual phoneme, 'iv': Niven, The Rivals, Livia. This may seem obscure, but in Kammerer's view many series contain repeated sub-elements. If you are aware of them, series are very frequent occurrences.

S4
I am leafing through a book of stories and pause to read one called 'Avocado Pears'. At the same moment my wife remarks: 'Look how huge the childrens' avocado plants are becoming.'

[Again, these series of two events are very common, and are usually dismissed as co-incidence. But such 'coincidences' appear more likely to occur between people who are emotionally close.]

S5
On my way to Vancouver for the day I hear a news report on the car radio about Trudeau, and this provokes thought about his possible successor, Mulroney, and the pretty name of his wife, **Mila**. Later, on my way to the university library, I turn on the radio again and someone is talking about a person from Montreal called **Lawrence**. Since I have a friend from Montreal called Lawrence, and I have been thinking of him the previous day, and furthermore he knows Trudeau, I think 'maybe there is a series here', and I deliberately start looking for another connected event – scanning street signs for a 'Lawrence'. [But these deliberate procedures do not, it seems, work in evoking series]. I arrive at the library with little time to spare, go into the stacks looking for books of poems I want to read by Louis Simpson and Keith Douglas. On the way along the shelves I am amazed at the huge area of books by D.H.**Lawrence**. I pick up a biography of Douglas and leaf through it. There is a picture of a girlfriend he had, called **Mila**. On the last page I read about his death in an explosion of a shell, in a battle in which German **.88** guns are mentioned. Then I open Simpson's poems at a page where the first line I see refers to the explosions of German **.88** guns. Later that evening, on my way home, I turn on the car radio. Someone is being interviewed called R.D.**Lawrence**.

S6

One day I have to interview two new patients, both women, who have called for appointments. (I have not had any new patients in more than a month.) Both turn out to be black, from the Caribbean (although from different islands) and both want to go into psychotherapy because of weight problems. (I have never had any black women patients before this).

S7

I get into my car, notice that the milometer reads 690096, and think vaguely 'it's the same each way.' Absent-mindedly, I put the car into gear and set off forward instead of backward as I intended: I have put the lever to Drive instead of Reverse. I have never done this before.

[This raises a basic question: is this a series in which the process forward / reverse is repeating itself, or am I unconsciously fulfilling it?]

S8

As I am changing my youngest daughter J's nappy she looks up at me and says: 'You got a head. Mummy got a head. I got a head. R [her sister] got a head.' I think, 'what a funny way of looking at us, a family of four heads. A few minutes later she comes to me with a new book to read, about four little puppies. On the first page is an illustration' of four disembodied puppy heads.

S9

My daughter R puts on a new yellow sweater, to come out with me to a printer's I have never been at before. I say: 'You look like a **canary**!' At the printer's the owner says he has a whole roomful of birds to show my daughter. They turn out to be **canaries**.

S10

Having returned from Canada to England and working in a hospital in Birmingham, I was walking down stairs with two assistants and said 'Jolly Good!' about something. (In my family, I enjoy using a facetiously abbreviated version of this: '**Jol**!') I said chattily, 'When I lived in Canada and said "**Jolly Good**" people just thought it was silly, so I stopped.' I waved goodbye to the assistants and walked out into the car park. The first number plate I saw had the registration M213 JOL.

S11
I decided to telephone a neuropsychologist (whom I did not know) who I had heard had written a handbook on memory, to see where I could get hold of it. I looked up the directory of psychologists and found he was based in Durham. In the same pile as the directory was a postcard of **Chichester** I had bought when down there some months previously for an interview for a job which I had decided finally not to take. I thought something like, 'If I had accepted that job and we had moved to **Chichester** at least I would have been able to use that card.' When I got through to Durham I was told that the psychologist had moved to **Chichester**. When I telephoned him there he told me he had just taken the job I had refused.

S12
I telephoned a small garage in Warwickshire early in the morning to arrange to get my car serviced. I could not recall the name of the boss. No answer to the phone anyway. In the car on the way to work I was listening to a Schubert Lied and when it ended the accompanist was named as **Edwin** Fisher. Of course, the garage boss's name was **Edwin**. Then the following morning I was wondering whether to telephone a trainee whose surname was **Hain** to change an appointment, and found myself idly thinking that **Hain** meant **Grove** in German. It must have been a Schubert week on the radio, as I found myself listening to one of his symphonies. At the end the conductor's name was given as **Grove**. [This is what Kammerer called a 'double series.']

Precognition series and event series
There is surely a difference in kind between a series such as S5, in which some of the events occur in my mind as anticipations of a name on bookshelves or on the radio, and series such as S6, where my two Caribbean patients form the series without any events in my mind. But in S6 there is at least a common factor: me, the psychotherapist both women want to see. Within

S5 there is a sub-series, of the German .88 guns, in which the two elements are both outside my mind and on the pages of separate books. But here too I form a link, in picking up the two books.

In other cases the mind in which one event occurs as an image may not be that of the person who registers the series. In S8, my small daughter makes the remark that we are a family of four heads, before four heads are seen in the book which she fetches for me to read (at random, and she does not know how to count). Although I register the series in my mind, its first event occurs in hers.

In S9 there are two separate events outside the mind, the yellow sweater and the room full of canaries. But there is also my remark that the pullover is 'like a canary'. Since this remark can be seen as precognitive in some sense, S9 might be shifted back to the Precognition series section of this chapter. But it remains in the Series section as an example of how precognition is not immediately apparent, and of how complex even a simple series can be. I propose, however, that any series in which one or more events occurs in the mind can be called a *precognition series.*

The second kind of series might be called an *event series*, in that no mental imagery or precognition occurs, it is simply a series of linked or repeated events occurring apart from the observer's mental processes until these processes are involved in registering them.

Kammerer's work in biology also relies on the delineation of what can be called *pattern series.*

How can we be sure that a series does not have a physical cause, even if this is hidden? Kammerer did his best to ferret out such causes, and often found them, although he did not think they necessarily disqualified the law of series. There is a *possible* cause for almost anything – as any evasive witness under cross examination (or a child being questioned by an angry parent) knows. Only in the last resort does the answer to the question 'Why?' have to be simply, 'Because!'

Nevertheless, many apparent series turn out to have such an obvious cause that it seems extreme to invoke 'the law of series'. One year, on the first bright day of spring, I found myself singing Shakespeare's song 'It was a lover and his lass…. Sweet lovers love the spring.' A few hours later I turned on the radio and someone was singing the song. But it *was* spring, and it is no series or example of extra-sensory perception that the radio programmer and I had the same idea. In an event series, the physical cause may be hidden but no less real. Three old men in a station waiting room carrying shepherds' crooks may all be going to a sheep-selling fair. Of course this can be investigated. Once, on my regular Vancouver Island ferry run, I found that there were no less than five mothers with small babies among the twenty or so people in the upper deck lounge. Not for weeks had I seen a mother and baby in this lounge which at this particular time of the week is full of businessmen. Had the spring weather brought the mothers out? Was there perhaps a mother and baby convention in Vancouver? No, apparently they had no purpose in common, and they did not know each other although they eventually got together and began comparing babies. And I doubt if my consciousness, which likes mothers and babies, is so powerful as to have somehow created the situation: if I had not been on the ferry that morning the rather unusual event would still have occurred, of five young mothers with babies finding themselves 'by chance' in the same place. According to Kammerer, such series are not at all rare, and an observer in any public place can note the occurrence of series after series. When on the look out for series on the street I have noted a series of blonde women with moustaches, a series of embracing lovers of whom the man was much shorter than the woman, and so on... It is just that we normally do not stop to look. When we suddenly notice a series we may be struck by it. And it is far fetched to claim that we are somehow imposing it on a random churning of events. Or is it that the world churns up such a huge variety of possibilities that some event clusters are bound to occur? This is the last refuge of the sceptic about series. But ever since Moire's 19th century best seller on *The Doctrine of Chance*, statistical arguments about probabilities outside laboratory conditions have been untestable. The world is simply too varied, and it is impossible to prove the null hypothesis – i.e. that the embracing lovers are *not* a series.

Many events do occur in clusters, as surely (Kammerer points out) as trains or buses tend to arrive at the stops in groups even if they have set out from the depot at separately timed intervals earlier in the day. Some of these coincidences may be explained, or explained away, by Per Bak's theory of 'critical events' – that they occur because of a build up of mechanical tensions toward a tipping point. But even people who live on quiet streets may notice how after long periods of no traffic, several vehicles will appear all at once – not simply because of the workings of some distant traffic light, but apparently from all directions. Kammerer explains such clusters not by invoking a hidden cause, as a sceptic might (the buses got into a traffic jam; there is a sale on baby clothes in Vancouver) but by invoking the *affinity* of events. Buses or bus-drivers in some way attract each other, as do young mothers, or lovers and ex-lovers as in my example P4.

Emotion in series

Even if all the young mothers admit to being drawn out by the weather, this may not be sufficient to explain the series completely. In a very striking series, such as this one, all of the 'logical', i.e. the causal, explanations which can be proposed still do not add up to total conviction. Something specifically strange and unexplainable remains, characteristic, it seems, merely of the series itself. The link is unknown.

I suspect the link is *emotional*, in all cases. Bus drivers or young mothers are on the same emotional 'wave length' as their companions or others with the same vocation. The striking timing of series P4 when our visitor passed into my work room at the same instant as her ex-lover telephoned was not a very emotional event for me, and they did not even know it occurred. Nevertheless they were, for the first time in over a year within a few miles of each other, as if within an unknown kind of emotional range.

In a precognition series, at least, the hidden links to the events may be at least partly in the individual's own mind, perhaps strictly speaking in the 'unconscious', though not in the Freudian sense. Neuroscience accepts that most processing of information occurs outside consciousness. If I see a green pasture and a few minutes later find myself humming Blake's Jerusalem, or if I look in a mirror and find myself suddenly humming the 'Tuba Mira' (the sound 'Mira' apparently evoked by the sound 'mirror') from Mozart's Requiem, there is a hidden linking process occurring in my unconscious. But if a series theory such as Kammerer's is applied to the absolute limit, then even such events could be seen as series. Why not just assume a series between green pastures and Jerusalem, between 'Mira' and 'mirror'? But this reduces us to mere spatial co-ordinates through which series weave their dance. And it is putting the cart before the horse, since series do not seem to be part of the flow of time: on the contrary, they seem to resist it, forms and patterns occurring in the general chaos, resisting entropy.

Some of these series are striking, but the emotional content is not always obvious. In the 'Lawrence' series (S4) I was consciously looking ahead for the series completion, but this was in a fairly calculating mood of curiosity. On the other hand the series contains the names of a friend who is geographically distant; a writer, D.H.Lawrence, whom I dislike but find disturbing because he attempts to resolve questions which I myself would like to resolve; a poet, Douglas, about whose work I am ambivalent, considering it alive but too forced, but whose death I find disturbing (and he fought in the same African campaign as my father).

The appearance of two Caribbean women in therapy was not destined to be emotionally neutral either: one of them I thoroughly disliked and she left therapy after a few sessions; the other I liked and I had the satisfactory experience of helping her and her husband resolve some problems.

I am also struck by how many series involve events which I experience with my wife and daughters. The element of emotional closeness, of attunement (or 'at one-ment') is marked in some series and in the 'more-than-coincidences' which traditionally occur between lovers. It is odd that 'like attracts like' – in both senses of the word. People who 'like' each other, and are presumably in some ways 'like' each 'other, seem to generate series of 'like' events.

Attunement, like that between the vibrating sound waves emanating from different musical instruments, also suggests a field phenomenon. If I look at the series I have experienced I have to conclude that their emotional content is merely a matter of degree. It ranges from startlingly intense, through the disturbing, to the quiet attunement of contact with someone I love. It is never neutral. This raises the question of whether *any* experience is really neutral: it may be proposed that every event evokes some emotion. I would deny this. I am not at all moved by thousands of phenomena of everyday life, particularly in a modern city. Nor do I seem to experience series involving such banal events as buying fuel for my car (although once when I was driving home anxiously 'bursting' with something I had to tell my wife, a rubber pipe in the cooling system burst);

or going into shops to buy food (although I might meet a person in a shop whose presence rounded out a series). And I would reject any pseudo-statistical explanation away to the effect that coincidences occur all the time and some of them I choose to notice for emotional reasons. I notice more-than-coincidences precisely because they are 'more-than'. Of course they may occur frequently and be impatiently ignored unless I am open to registering them – which means open to my emotions.

My own experience therefore suggests that *series which include one or more mental events are linked by emotion.* This can only be stated subjectively. There is no objective way of testing it. But it can be supported by inter-subjective checking.

4

SERIES AND TIME

Event Clusters

Dunne's claims about precognition in dreams have been tested on and off by respectable researchers (e.g. Ullman and Krippner, 1966–1969) and found to hold true, though to a lesser extent than he maintained. The 1970s seem to have been the most adventurous decade in terms of established scientists attempting to test various fringe theories. Funding has become tighter since then and such is the animus these theories provoke that conducting research on them is a fast way to losing respect and jobs. Research in ESP (extra-sensory-perception), even more disreputable, no matter who does it, is often linked to the theory of series. For example, if a person predicts what card will be turned up, this can be seen as a two step series: the thought of the card and its appearance. Even the CIA used to fund this research.

Card reading is supposedly a 'neutral' procedure, in which emotion plays no role. But perhaps putting out the effort to read cards, or opening one's mind to the reception of any image, is not always neutral. One 1970s ESP researcher, Puharich, claimed that autonomic excitation is a factor in all ESP, with the sender in a state of 'sympathetic' arousal (emotionally, tension), and the receiver in a state of 'parasympathetic' arousal (emotionally, relaxation). This is not true to my own experience. For example, Miss Fiona Love may have been relaxed on that train, but I – who was receiving the precognitive images – was in a state of anxious confusion.

Unfortunately, laboratory experiments with ESP may force the emotions, just as Masters' and Johnson's experiments force sexual excitement, or an actor may force tears.

Kammerer's theory of series ignores emotion (indeed emotion is Kammerer's blind spot, as time is Reich's) and concentrates on the objective record and classification of events. Nor did Kammerer discuss any theory of biological fields. Instead he sets out a set of theories of all sorts, a rag-bag of possibilities, as if he was jumping on every available bandwagon. (He was very ambitious). His position can be summed up as follows:

1) Series occur frequently and contain events or observations which have an affinity for each other.
2) These affinities can occur at all sorts of levels: in appearance, sound, language, patterns etc.
3) There is often a causal antecedent. E.g. you keep seeing bandsmen in uniform carrying instruments, and there is a band competition in town that day. But there is often no discernible antecedent, and the series seems to be acausal.
4) Kammerer 'cops out' of the rather interesting question of causality, which he seems to circle and never quite address: he flirts with the idea of an 'acausal principle' in the universe, but backs away from committing himself to it. In fact he resorts to a 'mystery' position, in which the causes of series will eventually be revealed by the laws of nature.
5) Series show an 'inertia', similar to that in Newtonian physics. Kammerer uses the word *Beharrung* which is the standard German translation of the Latin inertia, but which also suggests 'persistence' or (in military contexts) 'stubborn resistance'. Here he is suggesting that just as inertia permeates the universe, as the primary momentum of all bodies in it, the 'law of series' permeates events. So in spite of his flirtation with letting go of causality, he is now claiming that the law of series is a sort of primal cause.
6) The law of series can therefore be related to every phenomenon under the sun. Kammerer's book therefore consists partly of a systematic description of series, but mainly of a promotion of their importance. Here, Kammerer's Lamarckism is calling the shots: the law of series becomes an explanation (although a circular one) of how patterns of Lamarckian inheritance (e.g. the 'series' of spots on a lizards back) succeed each other.

Kammerer's theoretical conclusions eventually let his book down, and are a cop-out from several questions he has either hinted at or seems to have avoided. These can be summed up:

Is it in fact possible that series occur *acausally*? They just *are* or *occur*. Is there an 'acausal principle in the universe'? Do all things have to be caused?

Series amount to 'event clusters'. Kammerer does not use this term or any German equivalent, but Koestler does in his discussion of Kammerer's book, and it is a logical conclusion.

Do events cluster together because of an *affinity*? (Again, without actual causes.)

Do *patterns* – e.g. of the spots on a sole's back and the lumps in the sand it is lying on, or of a chameleon's colour and its background – also cluster together, through affinity, like events?

Is it possible that Kammerer has got his theory upside down? *Beharrung* derives from an old Lettish word meaning 'hope.' The 'persistence' of series in a universe of entropy – general running down – is rather like the persistence of life. (Kammerer hints at this, in a discussion of entropy which looks ahead to Schroedinger's proposition, some 35 years later, that life entails local 'negative entropy'). Series may not be an expression of inertia, but of something working against it.

Being a vitalist, Kammerer hints that series are linked with some sort of vital energy, but again backs off from this – which is a shame, in my view, because of the paradox that if he had begun to think of series as an expression of 'life energy' he might then have worked through this and realised that what may characterise series is that they require *no energy at all* – or, more accurately, that like all physical events they may involve a discharge of energy, but what makes them series is quite independent of energy: it is, again, *affinity*, which after all is a *relation*.

Behind all the above questions is the idea that series may express an *acausal* principle in the universe, one of affinity and relation. Series persist *against entropy*. They express a local order – as does life. They also, rather obviously, *persist against time*. They cause the observer to pause and reflect on the *co-incidence* or *more-than-coincidence* of affinities that are normally expected not to co-occur in clusters in time. It is as if the flow of time suddenly falters, and its elements congeal.

Series against time

If series persist against time, we can ask: against *which* time? Against spatialised time? Or against lived time? This distinction of the French physicist Henri Bergson's was a huge influence in the early 20th century. He proposed that time as measured in spatial terms (the movement of the hands of a clock, the trickle of sand in an hour glass, the progression of days and nights), *le temps spatiale*, was distinct from the subjective experience of duration (time seeming to run quickly or drag slowly) which he called *le temps vécu*. This led to a mindless vitalism that swept Europe. (The Italian poet Montale called it 'brute vitalism') As a student in Vienna after the First World War Reich described himself as 'a crazy Bergsonian'. But Bergsonianism had already failed in practical terms as thousands of French cavalry troops motivated by *élan vitale* – a sort of 'mind over matter' belief – were slaughtered by the spatialised machine-gun fire of the invading Germans.

Bergson, although a physicist, rejected spatial time and wrote almost endlessly about lived time. He saw it as a flow, a perpetual *flux*. This was the view of Parmenides' opposite number, as it were, Heraclitus. From the Bergonsian point of view, Parmenides is setting out spatial time, Heraclitus is setting out lived time. But the Heraclitean flux is, from the point of view of classical physics, which Bergson shared, the same thing as *entropy* – the running down of the universe in inexorable progress from order to disorder. Although later Schroedinger spelled out ways in which life could be 'negentropic' and involve local reversals of entropy, Bergson's picture was more dramatic, romantic, and mushy: as the forces of entropy 'descended' inexorably towards death, the life force (*élan vitale*) resisted and slowed this descent. As Wyndham Lewis pointed out in a famous attack on Bergson (whose lectures he had attended in 1912) the blind optimism with which he presents the life force is dragged down by the depressive dwelling on entropy and death. Finally, Lewis cracked, the *élan vitale* was just a parachute.

From a vitalistic point of view, it is all too easy to jump to a conclusion (though as usual Kammerer does not jump to it – he just goes up to it then backs off) that somehow series, or the perception of them, represent a force of life. At any rate, the subjects of series, coincidences, ESP, synchronicity, dream precognition etc. are all grist for the mill of mysticism. People who are interested in these areas of 'fringe science' are usually trying to stake out a territory for a religious view. From this point of view, series are clusters of events or patterns in lived time – not in the spatialised time of classical physics, with its ultimate risk, in the theory of space time, of abolishing time altogether.

But stepping back a little, it is obvious that *series exist as patterns in space.* E.g. the patterns on the lizard's back, or the patterns on a possible graph of the arrival of three red headed men on the same bus. Leaving aside Kammerer's theorising, and his vitalistic tendencies, he has provided us with a treasure trove of examples of series which are evidence that events and patterns in spatialised time are often organised by a principle of affinity – of likeness, and not by cause and effect.

Far from being evidence for vitalism, the law of series provides evidence against it. Series *persist against lived time.* (This is why we find them striking!).

Precognition and time

The question of precognition, as described by Dunne, leads to similar conclusions. Obviously precognition calls into question the assumptions of lived time, in which we expect one moment to succeed another according to the laws of life, in which organisms are born, grow, reproduce, and decay. Precognition also calls into question our projection of lived time onto the universe, which is born in the Big Bang, and is expanding constantly along an 'arrow of time'. For both ourselves as individuals, and the universe, we suppose that this moment is succeeded by the next moment which is predictable up to a point by the laws of physics and probability, but which certainly does not *now* exist.

When I *know* that Miss Love is called Fiona, before I see any evidence of this, and without any reason, and I am thus seeing ahead in time, then the arrow of time is no longer flying. One single experience of this kind is enough to refute the idea that whatever happens ahead of the present moment is unknown. *The future is accessible in the present.*

For a while my experience of Miss Love became a paradigm for me of how poetry and emotion do not obey 'the laws of physics' – meaning the physics I had been taught at school in the 1950s, which was approximately vintage 1910. But I was wrong. Precognition does not, indeed, obey the laws of a unidirectional time. But the laws of physics have moved on since Newton, and especially since the Copenhagen theory of quantum physics. There is no arrow of time. Everything has already happened. Perhaps it is unusual, but not impossible to see ahead in time – because it is in effect seeing ahead in place (not space), like being given a glimpse of a distant landscape.

Series and Cause

Our normal consciousness expects time to unfold, like a blossoming flower, or to fly like an arrow. Series, and in particular precognition (which is probably best seen as a sub-type of series) disrupt this expectation. Dunne proposed a theory of consciousness as a series of Observers (Observer 1 observed by Observer 2 observed by Observer 3, and so on) which he could not prevent, in spite of ingenious argument, from becoming an infinite regression.

If he had had access to the findings of modern neuroscience he might have formulated a more solid theory. For example, Damasio in his book *The Feeling of What Happens* postulates three levels of consciousness (the 'proto-self', the 'core consciousness' of any organism, and the 'extended consciousness' which only humans possess.) Damasio can make a coherent attempt to locate these levels of consciousness in different brain systems. But extended consciousness is an 'observer' of the simpler levels of consciousness. Dunne was not on the wrong track. Like Kammerer, he failed at the theoretical level, but not the observational level.

The arrow of time assumed in Newtonian physics was superseded in Einstein's relativity physics by 'space-time' – in effect 'block time', a four dimensional extension in space (time being added to the three spatial dimensions). It is in turn superseded, at least provisionally, in Julian Barbour's timeless universe of 'configuration space'. But assuming the block time Einsteinian universe, how can precognition occur?

One naive proposition is that just as organisms put out feelers, or extend their senses, into surrounding space, so do they in time. Perhaps we put out feelers along the time block toward the future, as an amoeba puts out pseudopodia into its spatial environment. Precognition could be the extension of pseudopodia in space-time. This would be consistent with the evidence for spontaneous precognition occurring with emotional excitement: the extension of pseudopodia in the amoeba, or of feelers in insects, is accompanied by physiological excitation. It is now broadly accepted in neuroscience (Libert, Cozolino) that an organism is already moving, by a fraction of a second, in response to a stimulus *or* from a spontaneous impulse, *before* the organism is aware / conscious in any way of what is happening. Consciousness lags behind action. Furthermore at higher levels of arousal in human beings emotion always over-rules thought. It is beginning to seem that we just tag along with what is already happening. Does this mean that in precognition we are already ahead of ourselves? Our pseudopodia have extended forward in time and we catch up with a *prise de conscience*, a sudden awareness? So is emotion (= *moving out* in Latin) just a sign that we have already moved?

Such questions are so confusing that they call into question block space-time. They may only be resolved if time and even movement are abandoned. Just think: in a Parmenidean plenum our existence in the future is already *there*. Our pseudopodia are already extended in that direction in space we call the future. We may have subjective feelings of duration, but we demonstrably exist in spatialised time – which may turn out to contain no such thing as time, to be simply space or, since as Parmenides pointed out (and Barbour and his colleagues are currently pursuing) there can be no such thing as *nothing,* simply *place.*

No matter what theory of time we adopt, the mere fact of precognition poses problems. This is why most physicists reject precognition out of hand as a fiction, or more politely as an illusion based on a misunderstanding of the laws of probability. One accepted instance of precognition is enough to refute both the Newtonian arrow-of-time theory and the Einsteinian space-time theory – as surely as the observation of one black swan refutes the theory that all swans are white. We are not supposed to see ahead in time. The laws do not permit it. But if instances of precognition accumulate and are documented, there is a case for review of the laws. Similarly I remember seeing one of my daughters smile within an hour of being born, and she smiled frequently from then on when she seemed contented. But at the time every single book on child development stated as dogma that infants did not smile until they were 3 or 4 months old, and that any mother who thought otherwise was deluding herself. The (behaviourist) dogma was that a child had to *learn* how to smile. Now, 25 years later, textbooks contain photographs taken via ultrasound which show foetuses smiling in the womb, a month or so before birth. The (neuroscientific) view is that the smile is an innate behaviour: it is 'hard wired' in the brain, although it may often take social experience to activate it. As the saying goes in neuropsychology (and elsewhere): 'Absence of evidence is not evidence of absence.' That not enough mothers were taking photographs of newborn infants smiling did not mean they did not smile. That not enough careful evidence of precognition has been presented or attended to does not mean it does not exist.

As a child of about 6 I was looking out of the window of our house in a Belfast suburb during a rainstorm. A ball of fire about the size of a football came floating down slowly through the rain just across the street, blazing with bright bluish-yellow flame (rather like the colour of the gas jets on our kitchen cooker). As it landed on the pavement it vanished. I thought I heard a 'phut!' sound through the sound of the rain. After a while when the rain eased I put on my raincoat and went across the road to look. I was expecting a scorch-mark – like one further down the road which remained from a German incendiary which had fallen there during the war – but there was nothing. When I told my parents about this my father said something like: 'It must have been a fire-ball. I've heard of them being seen in thunderstorms. Maybe there was electricity in the air and a fire-ball formed. But they're not supposed to exist. Anyway, now you've seen one, we know they do.'

This is what I meant when I wrote that my family were rational but they accepted the irrational. I am grateful. Fireballs of this sort are occasionally described but still not acknowledged to exist so far as I know. They are not the usual meteoric fireballs or 'bolides' which burn up miles above ground. Perhaps an explanation will be forthcoming in terms of plasma fields. I hope so. Because I did see the fireball. Again, absence of evidence is not evidence of absence. And I am signed up to the necessity of evidence and of a scientific approach to unusual observations. But of course, being unusual they are often not replicable. Peculiar atmospheric observations, including so-called UFO sightings and frogs dropping from the sky, are sometimes known as 'Fortean' events, after the sensational news reports of Charles Fort in the early 20th century. But Fort included terrestrial phenomena like human spontaneous combustion and made no claims to being scientific. Two neuroscientists Michael Persinger and Gyslaine Lafreniere wrote a daring book on *Space Time Transients and Unusual Events* (1977) in which they proposed that some 'unidentified atmospheric phenomena' ('UAP') involve interaction between electromagnetic fields in the atmosphere and in the brain.

Interestingly, Persinger and Lafreniere's theory is taken into account in a British Ministry of Defence study of reported sightings of so called UFOs done in the late 1990s and made public in 2006. The study also demonstrated that UAP tend to occur in clusters, geographically and in time. In other words, even singular, unreplicable observations can be investigated 'en masse' using a statistical approach. Perhaps the scientific investigation of series could benefit from such an approach. It would be fascinating to know where and when the 'event clusters' of series occur.

For me, however, the one-off Miss Love experience was enough. I can never forget that sense of my mind being several seconds (not long – but enough!) ahead of events. I hope it has not made me gullible. It has certainly led me to take my own experience seriously, whether or not it fits accepted theories.

Acausality

My other experiences of 'precognition series' have been less dramatic, but numerous. They have also been less personal. I suppose it could be argued that some form of subtle non-verbal communication (though not telepathy or ESP – even less acceptable than series) occurred between me and Miss Love, and even between me and the lounge-lizard on the train. Although I was smitten by Miss Love, she was certainly not smitten by me. But there was an interaction. That is why the less dramatic instances of series are convincing. Most do not involve interactions, and many involve the impersonal medium of the radio churning out information almost randomly. In the sort of emotionally meaningless series (S3) where the recurring element is the letter combination 'i v' it is hard to believe some quite neutral process is not at work. It is like a recurring pattern woven into events. But the imagery of patterns suggests a spatial web. Event clusters occur in the web – or the block, if that is what it is – of space-time. This is all very well if we are looking at past events. Any series appears to have a beginning (even if it is a precognition series and the beginning is a thought not an observation) and an end. But as a pattern is woven, some of it remains in the future until the end is reached. The same is true of series. How can a series patterned into block time already have a future when the first event occurs? We cannot establish causal links between the 1st and the 5th event of a 5 event series. A series is not like the making of a cup of tea where the arrival of the tea is entailed in the boiling of the kettle. Or like birth where a chain of causes and effects ultimately leads to death.

In a series there appear to be no causal links. (A logician might argue for 'lateral causality' but in many series the argument would be forced.) What are the causal links between the burning down of 3 shops owned by 3 bald men, in Belfast, Ottawa, and Ladysmith, Quebec, between 1949 and 1976? There are no causal links. The link between all elements of the series is the observer – me. And part of the series is a wholly non-rational and inexplicable dream precognition by me. But I am not creating or causing the 3 bald men and the dream. We are all, it seems, part of a pattern. I am balding too but I don't own a shop. If I did perhaps it would burn down. Thinking of which (literally, as I write this) I realise that the only time I have been the owner of a 'shop' it was a 'print-shop' (a term used for a printing workshop, not a sales-shop) which I set up as a book printer and publisher in Ladysmith in 1968, in a disused wooden house on my land. I moved away in 1976 and the property was sold. In 1979 or so the wooden house, the former 'print-shop', burned to the ground. So the series includes another event I had not thought of. I hope the realisation of this is the 'last word', as it were, and the pattern is complete.

Again, the series forms a huge pattern of associated (through me) but not causally related events. Of coincidence. Or more-than-coincidence. At one point in the series (the precognition dream) I realised that I was in a series. I anticipated the next event – but inaccurately, as I encountered Mr Devine, not a burning building. On the other hand I encountered him in front of a building that would burn. So perhaps there is an element of dream prophecy. (I remember my father talking about 'psychics' he had met, saying 'The trouble is their prophecies are only half right: they say something will happen next week but it happens next year and in a different way.') It seems that in some way my mind leapt ahead, in the dream, to an event in the coming day. If so this is a breach of the laws of time – meaning the laws of time as an arrow. But actually the events of the series are so spread out that time is the last thing to consider. Viewed retrospectively, the series is just a rather strange story. It has happened. It is 'history.' There is no need to consider time at all when discussing it. The events in the cluster are linked by affinities (balding men) and images (burning shops) but not by time.

When we make time links (infer time) we tend to think causally. 'That old store was built 50 years ago and of course those wooden buildings are fire traps. But Mr Devine loved it. How proud he was of those shiny hardwood floors! He offered an old fashioned kind of service and people trusted him. The new building will never be the same.' The narrative includes defined causal links. In fictional narratives where events are influenced by coincidences or non-events – such as when in Hardy's novel *Tess of the Durbervilles*, Tess's fiancé does not receive a crucial letter from her because she has slipped it under his door and it has slid under the carpet – some people complain that the plot is 'not realistic.' (Hardy, typically, thought that life *was* like that.)

We tend to impose time on our lives and on our environment. The whole bizarre story of the bald men and the burning shops may be irritating the reader by now. But it did happen. Its events are linked, related. But not by time. And not by space. And not physically by a person. (I am personally aware of all its events, but I was not present at all of them). So where and when did it happen? It consists, finally, of itself. It is a whole with related parts. The relations are its key elements. Not coordinates in space or in time. More-than-coincidence is timeless.

5

AWARENESS OF TIME

Definitions of Time

How would you know if time had stopped? What would you see? What would you feel?

As I write this I can glance above my computer screen and see a late afternoon London sky framed in my window, a February sky with blue patches and darkening yellowish grey clouds, and against them a black silhouette of roof tops and chimneys and skeletal trees. The tree branches are waving and quivering in a breeze, the clouds are inching sideways behind the branches. A few birds appear, black silhouettes, moving from left to right across the frame. I turn my attention to what I feel, and become aware of my breathing, and of the sporadic movements of my fingers on the keyboard. I have just shifted my awareness, or consciousness, from this text to the scene outside my window, then to my bodily sensations, and now back to the text.

I suppose if time had stopped before the last paragraph the text would not exist – and nor would I. Life seems to depend on time. Certainly if the world stopped, so would I. But what if I stop first? I assume that if I had died before the last paragraph, the tree silhouettes would still be moving against the fading light, and occasionally birds would fly across the sky. I am equating time with movement – the world's, and my own. This is presumably why common sense requires that Parmenides' vision of an unmoving universe implies a timeless universe.

Common sense associates time with movement – the 'passing' of time. But this does not necessarily mean time *is* movement. There are all sorts of movements. The clouds are moving with the wind, the light is fading as the cycle of the day proceeds, the birds fly across the sky in their own good time. I can look at my watch and see it is 5 o'clock, by a shared convention of time. My brother in Ireland can look at his watch at the same time and see it is 5 o'clock, but according to the convention my watch – my time – is more accurate than his because I am on the Greenwich meridian and he is

6 degrees West where the sun goes down 20 minutes later. Clearly there is an intricate relationship between time and movement, but if I lift my arm and scratch my head this has nothing to do with time – unless it is agreed to be so. I might have said to my daughter: 'When I scratch my head it's time to put on the kettle.'

So time is also associated with consciousness. To some extent we construct it. We can see it: the second hand of my watch keeps ticking around, and I can see the light fading in the sky. But we also feel it. I paid attention to my breathing.

Galileo, who lived in an age without reliable clocks, could not, in 1603 or so, use an hour-glass to measure how long it look an iron ball to roll down a chute in an experiment – since in an hour-glass time is, as it were, pre-measured. Instead he used his own body as a measure. He counted his own pulse beats.

If there were no life in the universe would time exist? Here all sorts of arguments can open up. Perhaps time is a human invention so it wouldn't exist without us. Perhaps time exists so long as the cyclic motion of the universe continues, and the whole universe is a cosmic clock. It all depends on how we define time.

Definitions of time are among the longest entries in any dictionary. They usually start by noting finite periods: 'A limited stretch or space of continued existence'. They then note 'Time when. A point of time.' And finally more generalised, indefinite meanings: 'Indefinite continuous duration regarded as that in which the sequence of events takes place'. These examples from the biggest Oxford English Dictionary suggest the size of the problem: there are times within time.

Taking the long view, different concepts of time are evident in the etymology of the various words for it in the Indo European languages. Carl Buck in his extraordinary compendium of synonyms from over 30 Indo-European languages immediately draws attention to the fact that 'Words for several of the notions classified under "Spatial Relations" or "Quantity and Number", like "long", "short", "first", "last", are applied equally to time.' For a couple of thousand years before Einstein's 'space time' was proposed, Indo-European languages showed a tendency to conflate the two concepts. Buck also notes that all Indo-European languages make the same distinction as English between 'time' as applied to specific periods and 'time' in the more general sense of a duration in which these periods are embedded. Ideas of time as 'stretch' in space, and as recurring cycles, occur in almost all European languages. An example of 'stretch' is the way in some languages the same term is used for 'time' and 'weather' (e.g. French 'temps', Irish 'aimsir'). An example of recurrence is the tautologous English phrase 'time and tide': 'tide' is the same word as German 'Zeit' ('time'), and the German word 'Zeitung' for a newspaper is a 'tiding'.

Clocks, hour glasses, and sundials 'measure the passage of time' but the distinction between the measure and the measured is not clear. Could it be that they measure themselves? What is happening when 5 minutes on the clock seems like an eternity, or an hour seems to pass in a flash – according to what we are feeling? To say the least there is a discrepancy between subjective time (our sense of duration) and objective time (the movement of the clock hand, the pouring sand, the shadow on the sundial). But then there is Galileo's pulse: he used himself as a clock.

There are as many potential clocks in the universe as there are recurring cycles. And cycles are correlated with each other. The great botanist Carl Linneus proposed a Flower Clock: '6 a.m. Spotted Cat's Ear opens; 7 a.m. African Marigold opens; 8 a.m. Mouse Ear Hawkweed opens…' and so on. Observably these cycles correlate with the position of the sun. But then the cycles will continue whether the sun is out or not, or for some time if the flowers are removed indoors, or flown to another time zone – although they will soon adjust. Many cycles appear to be completely internal – such as the deep sea crayfish which lets out a flash of light every twelve hours, but is never exposed to light. We have various hormonal clocks. There is increasing evidence that we have a 40 herz rhythm distributed throughout the brain. We even have molecular clocks. Universal time is now measured according to the vibration rate of the caesium atom.

So are we each our own clock? Each of us with our own time, or set of times – an intricate harmony of beating hearts, breathing lungs, hormonal tides, digestive cycles, sleeping and waking? I imagine most readers would answer this question with a cautious 'Yes, but…' We are aware of being less reliable than the watches we wear on our wrists, less regular, more variable. Like Galileo's pulse. We are not machines, we are not yet robots – or at least we hope not.

Pulsation and Pulse-Waves

[This and the following section about pulsation and oscillation also occur in my book, *Pulsation*.]

The word *pulse* is from Latin *pellere* – to beat. A pulse is a blow. We have come to use *pulse* for any regularly repeated phenomenon – such as the pulse of waves on a shore. Astronomers have discovered that quasars pulse – i.e. they flash regularly – like a light-house. But it is unkown whether the quasar is rotating or just flashing on and off. A distant observer cannot tell whether the pulse of light from a light-house is from the sort of light-house which switches a beam on and off at fixed intervals, or the sort in which a rotating beam covers a fixed point at intervals. The pulse of our heartbeat is also just an indication of other events: if we listen to our heartbeat on the pillow on a sleepless night we hear a double thump.

One of the fathers of scientific physiology is William Harvey, a contemporary of Galileo and 'the discoverer of the circulation of the blood'. 'Every schoolboy knows' that before Harvey the blood was simply assumed to fill the body, and the heartbeat was viewed as being a particular function of the heart itself, not the 'pump' which Harvey discovered it to be. The heart is a specialised machine 'for circulating the blood' (mechanistic science often favours this kind of creeping teleology) among organs which are also specialised machines. But this was *not* Harvey's view. Nor did he ever use the word pump or its Latin equivalent (he wrote in Latin). He did discover that as the heart contracted it sent blood spurting down the arteries, and that this caused the blood to circulate. This was, evidently, one of the principal bodily mechanisms. But in its turn it was a result, not the cause

Harvey wrote of watching the cloudy spot in a hen's egg four or five days after incubation:

> In the centre of the cloud there was a throbbing point of blood, so trifling that it disappeared on contraction and was lost to sight, while on relaxation it appeared again like a red pin-point. Throbbing between existence and non-existence, now visible, now invisible, it was the beginning of life.

The throbbing Harvey describes is the sign of alternating contraction and relaxation – as in the heart – and he is using it as a definition of life. Words like 'heart-throb' suggest that 'throb' is connected with the living body. We would not say a light-house beam throbs – or a quasar. So the throbbing of life is more than simply the regular appearance of a pulse or beat.

It is useful to get teleology – explanations in terms of purpose, which suggest design – out of physiological descriptions. Harvey did not write teleologically. He did not say or imply that the heart pulsed *in order to* circulate the blood. Rather, the whole body throbbed with the circulation in which the heart was the *main* mover.

In the late 20th century various 'New Age' thinkers (e.g. Fritjof Capra in his *The Tao of Physics*) describe the universe as full of oscillations and vibrations – but their function is not discussed. Even Rupert Sheldrake, in an innovative theory of formative causation based on 'morphogenetic fields', although he refers frequently to oscillation does not distinguish it from vibration, nor does he distinguish between its nature in the living and the non-living:

Atoms, molecules, crystals, organelles, cells, tissues, organs, and organisms are all made up of parts in ceaseless oscillation, and all have their own characteristic patterns of vibration and internal rhythm.

Part of the task of distinguishing life and non-life must be to discover if alternating expansion and contraction, commonly known as oscillation, is a valid criterion for life. If oscillation is as the *Penguin Dictionary of Physics* bluntly defines it, 'a vibration', then it is *not* characteristic of what common sense and ordinary language call life, although it may be considered to be so in the tautological worlds of animism or hylozoism, where everything is alive. In the mechanistic world of modern science, where everything is vibrating at different rates – including the hypothetical 'strings' which many physicists believe are the most basic units of matter / energy – everything is dead.

The way out of the confusion of oscillation and vibration in life and non-life is, I propose, to distinguish a particular kind of *phase unequal* alternating expansion and contraction as *pulsation*, and to claim that it is characteristic of life and life only.

This idea originates in Wilhem Reich's theory that pulsation is the life function. But Reich nevertheless made no distinction between pulsation and ordinary oscillation. In fact he grafted the word pulsation onto a theory first set out in terms of oscillation.

The current (online) Harcourt Academic Press Dictionary of Science and Technology offers definitions of pulsation under two headings:
'Pulsation. *Physiology.* A regular swelling and shrinking motion, such as that of the heart muscle.'
'Pulsation. *Astrophysics.* The swelling and shrinking of a star as it evolves from the main sequence to the red-giant stage.'
Although swelling/shrinking is more concrete in English than expansion/contraction (the words being Anglo-Saxon rather than Latin) the definitions still offer no distinction between life and non-life. Furthermore English dictionary definitions do not distinguish between pulsation and oscillation. According to the Oxford English Dictionary (OED), Pulsation is 'Rhythmical expansion and contraction; beating, throbbing, vibration.' But looked at closely these are distinct functions. Adding to the confusion is the OED definition of Oscillate: 'Swing to and fro. Vibrate.'
In contrast to the OED, the German Wahrig Dictionary provides a clear distinction between Pulsation and Oscillation (being technical terms the words are the same in German as in English).
Pulsation is: 'Activity of the heart; the consequently evident pressure-waves in the arterial vascular system' [Tätigkeit des Herzens; die dadurch erzeugten Druckwellen im arterielle Gefäss-system]. Oscillation is: 'Swinging: regular movements' [Schwingen: gleichmässige Bewegungen].
'Vascular system' is the normal translation of 'Gefäss-system' but literally *Gefäss* means a container or vessel. The definition makes it clear that pulsation occurs in a *contained* system and spells out that this is associated with *pressure*-waves. On the other hand oscillation is simply a swinging back and forth (uncontained – as for example a pendulum). Furthermore Wahrig spells out that oscillation is *regular* and furthermore with the nuance in German that *gleichmässig* means *of equal measure.*

While living in Quebec, from 1967 to 1976, I often observed the aurora borealis, in an arc of greenish 'curtains' or 'draperies' from the East around from North to West, reaching up 60 to 70 degrees, sometimes in the form of playing 'searchlights'. At the same time there were huge luminous blobs flickering and darting across the zenith ahead of the main lights, forming and re-forming, appearing and vanishing at a main point 15 degrees or so South of the zenith. Darting extensions of the auroral searchlights from the NE and NW were crossing the zenith to this main point and hooking in circles around the luminous patches as they formed. These extensions seemed to 'swoop' on the main point, now simultaneously from NE and NW in pairs hooking around each other and spiraling into the main point, appearing and vanishing almost instantly. These movements took on a definite rhythm: streamers of light from the auroral curtain rushing across the zenith and into the main point, vanishing, rushing in again, vanishing, rushing in, at the rate of about one of these 'pulses' per second.

The various streams of light began to merge and even out into broader streams obliterating the main point and becoming a 'mainstream' extending over the zenith from a double-dome of light in the North, each side of the dome seemingly pushed upward by or containing bright internal shafts of light. The mainstream contained rushing light in various wave forms moving from two opposing directions. This mainstream was about 20 degrees wide, stretching across the former main point. There were now some luminous patches in the sky to the South, but the North was relatively empty apart from a few greenish shafts of light.

It was now midnight. The mainstream had settled down into a narrower 'rope' of light, about 5 degrees wide and stationary, with a few waves: like a braided rope or silver-green river across the sky, and much brighter. It remained stationary for about 10 minutes while the Southern edge gradually took on a pink colour which then permeated and replaced the green.

At about 10 minutes after clock midnight, i.e. at 'true' midnight for longitude 75, the main point suddenly asserted itself again in a 'soft explosion', a sort of starburst or 'unfolding flower' effect, dividing the mainstream back into two again.

The streams now diffused quite quickly leaving a sky similar to that at the beginning of my observation, the main shafts of light to the North persisting but less bright, the zenith crossed by vague and diffuse streamers and luminous patches.

My impression was of a process of 1) alternating diffusion and fusion, 2) accelerating pulses, 3) merging 4) total fusion 5) 'burst' 6) diffusion. Although unusually bright and active, this auroral display contained the same sequence of events I had observed in other displays, in what is called the 'auroral substorm.'

In conventional descriptions of the aurora borealis the substorm is characterized as having two phases, of *expansion*, and *recovery*. *Pulsation* (although it must be remembered that in mechanistic physics the word pulsation is used interchangeably with oscillation) is noted as occurring around the main point or 'auroral corona.' The word *streaming* is used for 'apparent movements of regions of enhanced luminosity across pre-existing auroral forms.'

Human pulsation

Unlike the so called pulsation of the aurora, which is strictly speaking a series of equal phase atmospheric pulses, human pulsation is unequal phase.

A heartbeat can only be diagrammed in a sine curve if time is disregarded. Otherwise the heart's diastole and systole must include phase-unequal curves. An electrocardiogram (a direct recording of electrical impulses between "lead" points as the heart beats) traces 5 forms, known as P, Q, R, S and T (with sometimes a faint concluding wave known as U). P represents a build up of potential as a slowly rising curve which descends more quickly: it is phase unequal. It is followed by QRS which is a sharp spike (Q is the start, R the peak, and S the end) representing the mechanical release of tension in the "beat": it is phase equal. T is again more gentle: it is phase unequal.

Other directly recorded electrophysiological processes such as activity in the brain on electro-encephalograph register as phase unequal, although as rapid spikes and domes, not waves.

Breathing is observably a pulsation. Normal breathing *in* is always shorter than breathing *out*, wherever it is observed, in infants or adults – or in animals, such as cats or dogs. It can be willed (pushed 'mechanically') into a phase-equal or phase-reversed rhythm, but this cannot be sustained for long, and would revert to phase unequal in sleep. It may also, in anxiety states, show a phase-reversal (where the in breath takes longer than the out breath) although this leads to hyperventilation, and cannot be sustained: the person must stop or loses consciousness. In normal breathing the long phase, breathing out, goes with contraction of the ribcage, lungs and abdominal wall, but with expansion of the diaphragm which is its centre.

A study of pulsation in the body shows a series of interrelated levels, sometimes expanding and contracting reciprocally, which can be called pulsations within pulsations. The point is not that there is an absolute link between contraction (or expansion) and one side of the phase unequal process, but that phase inequality is characteristic of pulsation and of life. In this book, the words oscillation and vibration are used to describe phase equal (mechanical) processes, but *pulsation* describes *phase unequal expansion and contraction* visible when charted out as asymmetrical.

Another example is the pulsation of a jellyfish. (This was one of Reich's paradigms of pulsation, though he did not discuss it as phase unequal). In my observations, mainly of *Aequorea aequorea*, a small jellyfish which swarms in summer in the NW Pacific, phase inequality is marked and consistent. The main pulsation is in the 'curtain' which forms the jellyfish's circumference and from which the tentacles trail. As the curtain expands, water is pushed behind the jellyfish and it then drifts with the curtain expanded; the curtain then contracts abruptly and is gathered in before expansion is repeated. The expansion phase is approximately twice as long as the contraction.

One area of investigation which produces clear evidence of unequal phase pulsation is brain imaging, and in particular high intensity functional magnetic resonance imaging (fMRI). This investigation, known as BOLD (Blood Oxygenation Level Dependent) fMRI, is used to determine blood perfusion. The diagram of the haemodynamic response, i.e. of biological activity in the brain, shows unequal phase pulsation. In contrast the fMRI radio pulses show as equal phase.

Consciousness

According to David Stove, 'An announcement that a poetry-reading is about to take place will empty a room quicker than a water cannon.' I agree. Over a period of 40 years I only attended one poetry reading ever, out of politeness to a friend, though recently I have given readings of my own poems and find I enjoy it. I might now attend a reading by a friend, but otherwise I would still run a mile. An equally quick way to get me out of a room would be to announce a symposium on consciousness. Along with 'energy talk' this is one of the great dead ends of today. I find the word 'awareness' fit for purpose when discussing consciousness at its most simple level. Consciousness extends all the way from the awareness presumably felt by an amoeba reacting to a pin-prick to what is strictly speaking *self*-consciousness.

The most brilliant, though almost incredible book about consciousness written in the 20th century is surely Julian Jaynes' *The Origin of Consciousness in the Breakdown of the Bicameral Mind*. (I would stay in a room where he was discussing it). But it is in fact all about *self*-consciousness. Jaynes writes:

> Or introspect on when you last went swimming: I suspect you have an image of a seashore, lake or pool which is largely a retrospection, but when it comes to yourself swimming, lo! Like Nijinsky in his dance, you are seeing yourself swim, something that you have never observed at all! There is precious little of the actual sensations of swimming, the particular waterline across your face, the feel of the water against your skin, or to what extent your eyes were underwater as you turned your head to breathe.

This introspection does not work for me at all – or rather it works oppositely to Jaynes's. When I think of my most recent swim I see the pool, the high walls and windows of the building it is in, some people swimming. Then I feel myself swimming along one of the lanes, stretching my shoulder muscles in the crawl, feeling the cool water against my face, and so on. At no time do I see myself from outside. In fact I almost never see myself from outside. In memory or current consciousness I feel myself in my own skin, looking through my own eyes. In psychotherapy I have sometimes worked with people who have a habit of what I would call 'giving away their eyes', looking at themselves from outside as they walk down the street or as they talk to someone else at a party, as if they were in someone else's body. I see this as an escape from self awareness, an extreme self consciousness about the impression they make, and they are usually either rather insecure people or rather vain or 'narcissistic'. At any rate this is not a description of *my* consciousness. I therefore appear to lack what Jaynes calls the 'analogue I' – the image of myself as seen from outside. So by Jaynes's standard I am not conscious.

This misdefinition from the start, based on Jaynes's own introspection, undermines his central theory that until about 1,000 BC human beings were *not conscious*. Maybe they were like me. But I think I *am* conscious. I am just not usually *self*-conscious. However Jaynes is on to something in his 'bi-cameral' view of the mind, and I shall come back to this later.

In the early chapters of his book Jaynes provides abundant evidence that many human activities can take place without the necessity of any consciousness at all. All that is necessary is *reactivity* which he excludes from (self) consciousness. I would say, though, that reactivity does involve *awareness*. It might be argued that this does not always apply – for example when a person imperfectly under general anaesthetic reacts to a pin-prick without waking. But there is some evidence that even under general anaesthetic the person may be aware of what is going on, although the anaesthetic de-activates the storing of this in memory.

The neurology of consciousness is complicated. For example a brain injured person can pay attention to one thing but be conscious of (report perception of) another. As mentioned earlier, Antonio Damasio defines three levels of consciousness:

Proto-consciousness
Core consciousness
Extended consciousness.

Proto-consciousness or the 'proto-self' includes what we would normally call the 'unconscious' or 'subconscious'. Extended consciousness is our autobiographical memory, what we know of the world. Core consciousness links the other two, in experience, providing a sense of self. But only a sense. The neuropsychology of consciousness is like a hall of mirrors. Since consciousness cannot be located in the brain unless by speculating that it is located everywhere – in which case it must include what is going on the brain *un*consciously, as well as what is being paid attention to – it becomes elusive. Similarly the 'self'. Although neuropsychology has abandoned the behaviourist 'blank slate' view of the brain and now identifies various innate behavioural systems, it cannot locate the self any more than it can locate consciousness. (Jaynes explained this more clearly than Damasio). This may be partly a problem of language. As Jaak Panksepp remarks, 'our symbolic linguistic systems emerged only recently in human evolution, presumably for purposes other than the pursuit of science.' Non-scientists do not usually realise that the jargon of so much science is necessary as it invents its language of description, though clumsily. 'Consciousness' and 'self' cannot be observed or clearly defined. They are, as Damasio says, part of 'the feeling of what happens.'

Barbour in *The End of Time* repeatedly describes consciousness as a 'mystery' and he and others (including Stove – his opposite in most respects) have pointed out that a physical explanation of the universe that does not include consciousness is obviously incomplete. Furthermore consciousness must be included in any explanation of time.

My own definition of levels of consciousness, based inevitably on my own introspection as well as on the neuropsychology, is:

Perceptual consciousness: 'awareness'.
Reflective consciousness
Trance consciousness
Self consciousness.

Perceptual consciousness is clear enough: consciousness as a function of perception. I suspect all organisms have it.

Reflective consciousness is, as the word reflective implies, the back and forth process of trial and error thinking. This must vary in complexity, but even the most simple organisms work by trial and error and I propose that in however simple a way they must experience 'reflection' as distinct from direct awareness of perception.

Trance consciousness will be described in the next chapter, on poetry, in terms of 'inspiration.' It has certain qualities of automatism in that the person is not directing the attention voluntarily and feels flooded by awareness at many levels at once. It can be mislabelled (by Jaynes, for example) as *un*conscious. But I would describe it as the opposite, as 'never more conscious.'

Self consciousness is as described above, Jaynes's 'analogue I' in which the person steps out of their perceptual skin, as it were, and observes himself or herself from outside. It is human, not animal, and it may depend (as Jaynes convincingly explains) on the development of language.

The demarcation of consciousness

In this chapter I am mainly going to discuss the most basic level of consciousness: 'awareness.' Being a Germanic / Anglo-Saxon word rather than a Latin one it has not been picked up in scientific jargon. But we know empirically what it means. It will do.

I take a similar approach to memory. Where possible in this discussion I call it 'mind'. For me consciousness and memory have become mainly clinical terms (as in 'loss of' or 'disturbance of' either) and I hope to refresh my thinking with more down to earth terms. On the other hand discussing other writers' use of the words 'consciousness' and 'memory' I shall use the terms.

Before leaving Damasio's delineation of three levels of consciousness it is worth noting that they are not *demarcated*. Their boundaries are blurred. In fact a 'flow' of information in the neural network is assumed to link all three: unconscious information from the biological inside from level one to two, conscious information from the social outside from level three to two, and poor old two permeated by both flows. The word *flow* is either used or implied in much neuropsychology. Even in the everyday world we talk about information *flow*. In the brain, information flows along the threads and tubes of the neural network of axons and neurones, aided by chemical neurotransmitters. Hence *neuro*logy and *neuro*psychology. But the neurophysiologist Candace Pert shows in The Molecules of

Emotion how only about 5% of information transmitted in the brain is via neurotransmitters. The rest is transmitted via neuropeptides. There are 101 of these at latest count and the list is growing.

Neuropeptides are in fact vesicles, granules, in the fluid of the brain. Each is associated with a particular process or processes. Better known ones are oxytocin, the 'care' neuropeptide involved in the letting down reflex of breast-feeding mothers and in the afterglow (for both sexes) of orgasm; prolactin (again maternal feelings); insulin (energy balance and regulation); ACTH (stress); and vasopressin (male sexual arousal). Less known ones are galanine (memory); Substance P (pain, anger); and neuropeptide Y (involved in circadian rhythms).

Neurotransmitters too are vesicles – smaller and denser than neuropeptides. Both appear in the neurons (nerve cells) of the brain, the neuropeptides throughout and the neurotransmitters nearer the synapses (the couplings of the axon from one cell and dendrites from others). Both neurotransmitters and neuropeptides are released into the interstitial fluid in which the brain cells, and the brain itself, float. They are the micro-organisms of a mini-ocean. If Pert is right and 5% of information in the brain travels via neurotransmitters, 95% or so travels via what the neurologist Richard Cytowic has called 'volume transmission'.

Awareness of time

Since Einstein, physics has jettisoned 'flow' models of time. (Bergson did not, but physics has subsequently jettisoned *him*). But if time exists in the brain, then it is associated with a visible and measurable information flow. Time estimates – 'how long does it take to walk from here to your house?' or 'how long have we been talking?' – call on the brain's store of information about time and its current experience of it. Accurate time estimates require intact brain functioning at all levels. A simple summary is that the sub-cortical areas register the 'gut feeling' of the passage of time, the cortical processing areas calculate it, the cortical executive control areas judge the accuracy of the estimate. The brain is saturated with time. According to surgeons, it pulsates according to several rhythms, mainly from the beat of the circulation of blood from the heart, but also with every breath. (There are about five heartbeats to a breath). And it contains many 'timers.'

The circulatory pulsation of the brain is probably its main timer. Then there is the breathing. Neuropeptide Y and the hormone melatonin are released according to circadian rhythms. And as Cytowic describes it,

The limbic system performs calculations at an internal cycles-per-second rate of 400 Hz but is governed by an outer clock of 5 Hz, the rate of the theta thythm. In other words, a high-speed calculator is embodied in a low-speed clock. The cortex also performs high-speed modular transformations governed by a low-speed clock of 10 Hz, the frequency of the alpha rhythm.

The alpha and theta rhythms appear as pulsations (unequally jagged traces) on the electroencephalogram (EEG). Their interrelation has a function. According to Cytowic, 'The state of the world is pumped into the cortex and an evaluation comes out one fifth of a second later, yet elements *inside* the limbic system are cycling furiously 400 times a second to carry out the intermediate steps needed to derive that evaluation.' But the limbic system is the *emotional* processing system of the brain. The interacting systems of time registration in the brain are ordered by emotion. Cytowic, like Panksepp, states the 'primacy of emotion.' The 'top-down' thinking of cognitive neuropsychology (which to give it credit, successfully overturned the 'black box' no-mind non-thinking of behaviourism) is being replaced by a neuropsychology in which most brain functions derive from emotion – including the evaluation of time.

Cytowic also identifies a timer, of 'electrical potentials that oscillate at 40-Hz', in attention. He states that

the time scale of the 40-Hz oscillations corresponds to the psychological time scale of attentional shifts from object to object. Neurons stay phase-locked for several hundred milliseconds, long enough to possibly establish and break their temporary functional configurations in roughly the same time frame that an individual's conscious attention moves from one physical object to the next.

This is perhaps the brain process that enables Julian Barbour's timeless universe of 'nows.'

If the brain is floating in and saturated by a mini-ocean of time, does this re-instate the concept of time as flow? Not if the flow is considered, as it was in the Newtonian model of classical physics, to be uni-directional – like a wave advancing into a void. As in the observation of a drop of fluid under a microscope, the flow in the brain is in all directions. And the flow itself is *not* time. Rather, the membranous bodies – cells, vesicles, blood vessels, the brain itself in its dural sack – are either suspended or fixed points in the interstitial fluid, or moving around in it like the unicells in a microscopic slide drop, and each is a source of time. As Leibniz put it when discussing the units of interaction he called 'monads', 'But for all this, it must not be said that each portion of matter is animated, just as we do not say that a pond full of fishes is an animated body, although a fish is.'

As Wyndham Lewis remarks, Leibniz here 'contradicts the average space-timer of post-Relativity philosophy, for whom *the pond* too is virtually organic.' Lewis might have been anticipating *The Tao of Physics*.

To repeat, the flow itself is *not* time. But it is teeming with time-sources, their signals coming from all directions. Out in the open, as it were under the aurora borealis, these time-sources are the sine waves or regular oscillations which appear to fill the universe as far as we can register them, coming from all directions, all happening at once. Within the brain, the time sources must relate to the universal sine waves – which are visibly present in the brain as the 'Brownian movement' of all particles in any liquid. Within the universe, Barbour proposes, time is not absolute but relative (and therefore is an illusion). Within the brain, the most complex system we know of in the universe, time is also not absolute but relative. All those *times* (plural)! But we do have a *sense* of time, and if it is an illusion we nevertheless have invented it, presumably out of the complex interactions of pulsations within the brain which contribute to our awareness of 'time'. But to go back to the image of the fish in the pond, the fish are aware of time, the pond is not. The pond or the ocean may flow. But it is organisms within the flow who are aware of the flow. Rather than to say 'we are conscious of time' it might be more accurate to say 'we are aware of the tides.'

Flow and resistance
It is also worth noting that the brain is full of structures, not only the lobes and cell columns and neurons of the brain tissue itself, but the busy little vesicles of neurotransmitters and neuropeptides, each with its own membrane. The vesicles do not appear to 'go with the flow' either. They must *resist* the flow as they are attracted or repulsed in chemical or electrical fields. I imagine the schools of tiny jellyfish, *aequora aequora* I used to watch in Pacific Coast harbours. They are sometimes forced to drift with the tide or currents but even then their ceaseless pulsation propels them in other directions. They are not strong enough to swim against most currents, but they appear to *resist* them. They have membranes, and a neuronal fibre which indicates sense (of a sort) and if they have awareness, surely it is their senses that provide it. An artificially designated unit – say 10cc – of interstitial fluid or of sea-water cannot be *aware*. It must have separate existence, its own membrane, its own boundary, to be aware. But once it fulfils these conditions it is likely to be not only aware but alive. It is a pulsating organism.

Awareness and pulsation appear to be coupled. We cannot, in ourselves, conceive them as being apart. At death they both stop, we assume, with the last breath.

Returning to the philosophical problem of consciousness, if at its simplest we equate it with awareness, then it is inseparable from pulsation. This eliminates the need for a theory of *emergent* consciousness, as has been frequently proposed, most notably by Popper and Eccles in *The Self and Its Brain*. The philosopher Colin McGinn has despatched this theory with elegant ferocity, arguing that mind cannot *emerge* from matter. But it can be *in* matter. It might be argued that a pulsating micro-organism or jellyfish, though alive and aware, does not contain mind. I think it does.

Kammerer claimed that forms in nature were expressions of series. One of his examples was the way in which leaves match each other in pairs or groups as they are positioned on a branch. Wilhelm Reich also used the example of leaves on a branch in postulating that the flow of life energy continually splits into two streams, which in turn split. (Both Kammerer and Reich seem to have read something of Goethe's discussions of the distribution of leaflets along a stalk expressing the unfolding of the whole plant). For Kammerer, leaves grow in series; for Reich they grow where the energy flow along the branch has divided. Kammerer's theory is more finalistic: it is as if the leaves (like other biological patterns) have taken their rightful place. Reich does not explain why the energy flow should stop and become a leaf rather than keep dividing into progressively finer branches; but elsewhere he describes organic shapes (beans, eggs, leaves) as occurring where 'the energy' begins to turn back on itself, forming a membrane or within one. Generally, Kammerer discusses resistance (the inertial persistence of series) and not flow. Reich discusses flow (of the supposed energy) but not resistance which in his work is associated with the human 'armour' against the movement of the energy and is only seen as having a positive or essential function when discussed as a membrane in organisms. (Reich was blind to the role of resistance in pulsation, equating it psychoanalytically with the resistance, to emotion and to sexual drives, which he called 'armour'. He announced: 'When I see armour I want to smash it.' No wonder his theory deteriorated in some hands into a mandatory 'Go with the flow.')

Hydraulic mechanics demonstrates that resistance produces splitting. A simple example is forcing a lemon against a knife, so that the lemon is divided. But multiple splitting can he observed in something as simple as letting an ink drop fall into a glass of water: the drop splits into two curved lines each of which forms a toroid (doughnut shape) at the leading end, which in turn splits at right angles to the original split and forms new toroids – and so on. The splitting continues for as long as space and quantity of ink allow, so that before the ink dissolves an inverted tree-like 'spray' is visible in the water.

Different resistances produce variations in the splitting. A boulder placed dead centre of a stream will force the stream to divide equally around it. If placed off centre it will force most of the stream around one side, and the resulting inequality will lead eventually to a curving of the stream bed, the presence of eddies, a deeper channel on one side, and so on. Resistances to a stream can cause it to split, sometimes symmetrically, sometimes asymmetrically. Whatever is swept along on the flow may become progressively separated. If a seaweed clump presents a symmetrical face to a wave and is broken, perhaps its fragments will be thrown up on the shore by the same or a succeeding wave (synchronicity); perhaps they will turn up at intervals (seriality); or perhaps they will be so far separated that no series can be discerned.

All this hydraulic imagery suggests that if time were a fluid then ultimately its laws might be established through some equivalent of fluid dynamics. And if every series is in fact traceable to some original event, then a fluid time, like Bergson's, might be acceptable to science, or even made compatible, in some new synthesis, with the block time model. But some series are not traceable to an original event. The precognition series is like a leaf pattern without a tree, or like flotsam and jetsam on the shore without a common source: suddenly on the beach there are three bright purple objects – a piece of seaweed, a plastic boot, and a fragment of plastic. In such a case there are two links: the attribute in common (purple colour) and the timing of their arrival. There is, as it were, a partial affinity in spatial terms (light reflecting characteristics are similar, but not shape) and a partial affinity in temporal terms (they arrive approximately at the same time). Phrased differently: in spite of their temporal and spatial affinities the objects remain *unique*, as ultimately, everything is. Aggregates, classifications, and analogies are all convenient averagings, but it is well known that no single leaf and no single twin (even) is the same as another. 'Identity' is a convenient theory, not a fact. (Leibniz postulated that it was logically impossible.) Nevertheless, because of their affinities we can describe the objects as clustered in a series in space and in time. Perhaps one clue to the functioning of affinities is the possibility that if an object has a spatial similarity to another object it will also have a temporal similarity: two similar leaves, for example, might tend to come out on similar days in spring. But of course their relation to space and time can be explained mechanically as 'the product of history.'

We invent time

The analogy of time beating against the resistance of consciousness like waves against a shore, its waves casting up on our awareness the consequences of historical events out of range of our perception, cannot hold if we realise the full consequences of a time flow theory. We cannot, in fact, function as a shore to time, or even as a rock around which it flows, because that would mean we were there first! The leading edge of a flow of time cannot encounter a pre-existing object. Admittedly, waves can form a beach by pushing up sand, but there has originally been some kind of shore for the wave to break against: the shore pre-exists the breakers. But nothing can pre-exist the flow of time, by definition: each moment of time occurs at its leading edge. Bergson and the other vitalists of the 'time cult' simply got it wrong. It is not the flow (whether absolute or relative motion) of events that creates time. It is not even our resistance *to* the flow – which requires that we pre-exist the flow, an impossibility. But within the flow we pulsate, and we register the differences between our pulsation and the wave pulses of the flow. And so we invent time.

We cannot imagine nothingness in itself, or the leading edge of a stream flowing into nothingness. But for the sake of argument and perspective we can imagine nothingness as the traditional 'tabula rasa' or 'blank slate', in this case as a flat and completely featureless beach over which the leading waters of fluid time are advancing. On an actual beach when the tide is coming in, the waves pound and retreat leaving a shallow eddying sheet of water whose leading edge probes forward in rivulets and wavelets, until the sheet is swept away by the next wave, and so on. The leading edge of time, if it is in some sense a flow containing dynamic movement is not, therefore, a monotonous straight line. It may be exactly 7.15 p.m. in this room and it is 'the same time' across the street, and in New Delhi it is 1.15 tomorrow morning and still 'the same moment'. But in none of these places is the same thing happening, either quantitatively or qualitatively. Time may be being lived more turbulently in one place than another, and even objectively it may be more turbulent: more events may be occurring within the same space.

But once we start to talk like this, we are not talking about time as we usually understand it at all. If time is an ocean-like substance, what distinguishes it from whatever objects or creatures it contains? Are we swimming in a sea of time – or are we as much part of 'time' as the sea? In this case the whole universe becomes 'time'. If so, then we might as well scrub the word 'time' altogether. There is no time: the universe simply *is*. Interestingly, this model can apply for both a space-time universe and a timeless universe. It is the 'essentialist' problem again. If the universe is *all time* or *no time* then nothing is being said.

But if there is a distinction between a part of the universe that perceives time, and a part that is perceived as time, then we are in business. The evidence discussed so far in this book suggests that the living parts of the universe invent time in their interaction with the non-living parts.

Lived and spatialised time

Bergson's distinction between lived time and spatialised time is not completely useless. Lived time is time lived *by the percipient* of spatialised time. Spatialised time is the ever perceivable oscillation and pulse waves that fill the world around us. Actually, since classical dynamics and physics are now dead, and if Barbour is right the oscillations of spatialised time are 'pure shape' – corrugations in the sand of the universe, not moving waves – then spatialised time is an illusion. We can only discuss shapes.

It is fairly easy to accept that what Bergson called lived time varies from person to person, just as it does from moment to moment. Some experiences seem to pass quickly, while others drag slowly. Time *here*, in this enclosed membrane which I call my self (some languages have no word for 'I' but use the word for *here*), is being experienced differently from time *there*, meaning in another person's place. Although of course if we are 'on the same wavelength' our experiences of time may be more similar, and if we are acting or moving together, or 'entrained', it may seem 'the same.' Performing a synchronous action with a person *not* on the same wavelength may produce no such sense of sharing the experience of time. If two lovers are incompatible in bed, one may be merely 'waiting for it to be over with' while the other is 'enjoying every minute of it.'

That spatialized time varies may seem more complicated. But surely even objectively what we call 'the passage of time' is much different in a square mile of Calcutta from in a square mile of the Sahara desert, or in a greenhouse and a few square yards of Antarctica. And biological research (see Luce's *Biological Rhythms*) demonstrates that when a person's lived time varies subjectively, his or her metabolism also varies.

We experience the rhythm of each day at least, even if we become so tired we forget the date. Spatialized time can be let go of, but lived time cannot, at least before death. As we live time we also resist it. This paradox of being with a process but at the same time against it seems to have been acknowledged since primitive times, or at least language suggests so. In Old English the word 'with' meant both 'along with' as it does today, and 'against.'

Resistance *within* the supposed time flow was seen by Bergson in confused and mystical terms as a tendency for life to reach back on itself, towards the source of the flow, to strive upward towards this source (God?) against the downward flow of time and matter – but ultimately to fail. Hence Lewis' wicked image of the élan vitale as a parachute. The poet Robert Frost, influenced by Bergson and by William James, was able to express this more clearly in 'West-running Brook':

> The black stream, catching on a sunken rock,
> Flung backward on itself in one white wave,
> And the white water rode the black forever…
>
> It flows between us, over us, and with us.
> And it is time, strength, tone, light, life and love –
> And even substance lapsing unsubstantial;
> The universal cataract of death
> That spends to nothingness – and unresisted,
> Save by some strange resistance in itself,
> Not just a swerving, but a throwing back,
> As if regret were in it and were sacred.
> It has this throwing backward on itself
> So that the fall of most of it is always
> Raising a little, sending up a little.
> Our life runs down in sending up the clock.
> The brook runs down in sending up our life.
> The sun runs down in sending up the brook.
> And there is something sending up the sun.
> It is this backward motion toward the source,
> Against the stream, that most we see ourselves in…

It took a poet, rather than a philosopher using 'poetic' imagery, to pursue the imagery to its conclusion, much more harsh, more clearly dualistic than Bergson allowed it to be, in which the stream of time is a 'cataract of death.' Frost sees that 'this backward motion towards the source, / Against the stream' is not going to prevail in the stream he observes. The world out there is as bleak as Barbour's Platonia where the possibilities for life are slim. But, 'It is this backward motion...*that most we see ourselves in.*' Whether or not the backward motion prevails out there, it persists in us – in our pulsation.

'The rhythmic beat of consciousness'

Self-consciousness can impede the flow of human actions. When I first learned to skate, at the age of 33, and was picking my way cautiously but enjoyably along a frozen canal, a woman and a little boy came skating by, and I heard her saying to him encouragingly: 'Look, there's somebody who skates even worse than you!' I immediately fell down. Similarly in sex, phenomena analogous to my falling abruptly on the ice can occur. Conversely we may experience ecstasy when merged with oceanic or cosmic feelings through contact with nature or with a beloved. 'Ecstasy' means being put 'out of place': there are no boundaries. Some people mystically long for a universal consciousness, transcending the boundaries of human individuality. Or not so mystically: as the painter Max Beckmann wrote, 'the sports maniac is the soul of the collective man.' Perhaps this wish to become completely merged into a whole is a result of self-consciousness, of being caught in the trap of character armour and yearning to break free. Perhaps we break free at death, but not necessarily into new consciousness if consciousness in fact originates in boundaries. While alive we may feel that our consciousness even of the cosmos depends on our sense of our own physical experience: ecstasy cannot occur without a place to stand outside of. John Donne knew this in his famous poem 'The Ecstasie', about the two lovers sitting on a grassy bank all day, their hands clasped, looking into each other's eyes. Love 'interinanimates' their two souls and makes them one. But

> We then, who are this new soul, know
> Of what we are composed and made,
> For, th' Atomies of which we grow
> Are souls, whom no change can invade.
> But O alas, so long, so far

> Our bodies why do we forbear?
> They are ours, though they are not we, We are
> The intelligences, they the spheres.
> We owe them thanks, because they thus
> Did us, to us, at first convey,
> Yielded their forces, sense, to us,
> Nor are dross to us, but allay.

'Allay' means 'alloy', the melting of two metals together. Donne maintains that whatever melting together of the two souls is occurring, it is also occurring in the body. He concludes,

> And if some lover, such as we,
> Have heard this dialogue of one,
> Let him still mark us, he shall see
> Small change, when we'are to bodies gone.

In more modern English: another lover seeing the two come back to their bodily selves will notice little change, their bodies as much at one as their souls have been.

The lovers instinctively know that too much resistance blocks the flow completely and eliminates sensation. But they also know that yielding instantly to the first rush of sensation means the excitation (as intensified consciousness) cannot build and pleasure is ephemeral.

Some of Reich's successors have gone so far as to claim that a great deal of sensation during sex is a sign of blocking the progress toward the convulsions of orgasm, and the more full the orgasm the less the sensation. This is a Puritan ideal of mechanical efficiency, more grotesque than Reich himself could have foreseen, a reductio ad absurdum of his emphasis on letting go fully to autonomic movement. But its opposite contains a grain of truth: too quick a letting go to movement does diminish sensation. Presumably in the interest of maximum efficiency Donne and his beloved should have rushed to copulate briskly at the beginning of the day, then spent the rest of it in socially productive work. Another poet, Graves, wrote of the 'honest first reluctance to agree'. Shakespeare wrote of 'lust in action' rushing too quickly to its conclusion: 'the expense of spirit in a waste of shame.' Poets may notice these things acutely. But they are common sense.

If lovers are carried away on waves of desire they still maintain their own rhythms for as long as possible, or try to be so close to each other that their two individual flows combine in a powerful joint assertion of a new found self against the flux. Or an opposite approach may be taken: rather than resist the stream, they move ahead of it. As Andrew Marvell put it: 'though we cannot make our Sun / Stand still, yet we can make him run.' In either case lovers persevere against mere automatism until finally they are swept away by it. Even in orgasm people tend to utter their lover's name as if asserting their uniqueness, their simultaneous surrender to autonomic pulsation.

All this is more than metaphor. Our most intense experiences illuminate natural processes. The sexual act is an example of how consciousness is the more intense, the more there is this simultaneous opening to the flow of sensation and resistance to it. Laura Riding, who wrote of 'the rhythmic beat of consciousness' also had the insight (or 'outsight', as she would have called it) that intense consciousness depends on resistance. She wrote of 'the mob fear of the organized society of time against those incorruptible individuals who might reveal life to be an anarchy whose only order is a blind persistence. In the energy of this persistence occur intense flashes, the poetry or lightning of sense.' She saw individual consciousness as resistance to *time*. This is, although the ascetic Riding would not have liked the comparison, similar to the intensity of orgasm for a couple who synchronously give in to the beat of the autonomic flow while synchronously and passionately resisting it: it is their joint insistence in being themselves which enables their consciousness to double as they are overwhelmed by the pulsation of orgasm.

The pre-occupation with time which led to the first draft of this book in 1985 was prefigured by poems. I wrote about time in 1984 in a poem called '&':

> And all we set against it is our skin,
> Stretched and battered like a tympanum
> By its relentless wave-pulse, so intense
> We clutch each other fast as we let go,
> All trace in us dissolved of present tense.

But I see now that the 'relentless wave-pulse' is not in itself time. The universe is full of wave-pulses in all directions and probably at all frequencies, without any absolute direction. It is when the wave-pulses encounter the tympanum of our membranes as enclosed organisms – our pulsatory resistance – that they become time. Then we tame it to the oscillations of pendulums or the vibrations of a caesium atom, and make clocks.

6

MIND

Unconscious Memory
The Victorian thinker Samuel Butler, identifying what he saw as a gap in Darwin's evolutionary theory, proposed in two books, *Unconscious Memory* and *Life and Habit*, that an organism's experience leads to habits some of which are passed on to succeeding generations in the form of automatisms, behaviours of which the organism does not have to be aware. Hence the term 'unconscious memory'. The idea of the 'unconscious' was not uncommon at the time. (It was certainly not invented by Freud. It had begun with Leibniz in the 18th century). Darwin's cousin Francis Galton proposed that many thoughts and memories were suppressed or stored away from consciousness in a sort of cellar of the mind. Freud's theory of 'repression' added the imaginative twist that such information was not only stored out of consciousness, it was also unconsciously stored. We did not know what we had suppressed. 'Repression' developed the special sense of storing away sexually forbidden material which could only be retrieved, conveniently for Freud's power-play, by being interpreted by a 'psychoanalyst'. The British psychologist W H Rivers convincingly refuted this concept in 1921 or so and suggested a return to the more accurate term 'suppression' – i.e. the conscious storing away of painful material in the unconscious, which the person could eventually retrieve voluntarily, with emotional support, although he or she might prefer not to since this involved re-living the suppressed pain. Rivers' friend Robert Graves made the common sense point (in *Poetic Unreason*) that 'repression' works both ways. As he put it, Mr Hyde can repress the memory of Dr Jekyll as much as Dr Jekyll can repress that of Mr Hyde. But by this time stopping the express train of psychoanalysis was impossible.

For most of the 20th century the psychoanalytic view of memory as largely unconscious co-existed in a hostile stand-off with behaviourism which saw memory simply as the engraving of information on the blank slate of the brain. In behaviourism memory consisted of information which was registered by the senses, held in short term storage in the brain then in long term storage, and retrieved through free recall or recognition recall. As in computer science: 'information in, information out.' The brain was considered no more alive than a 'black box.' The first information to enter the memory of a human being or other organism was assumed to depend on the readiness of the senses to receive it. In the case of humans, this was assumed to occur in infancy. Most of us have a 'first memory' dating back to age three or four. Meanwhile the psychoanalysts worked diligently to 'make the unconscious conscious' and various pre-verbal memories from earlier in infancy were dredged up and verbalised in analysis. Then behaviourist research demonstrated that the foetus could learn from about the 7th month in the womb, and respond to different stimuli (e.g. sounds which pass through the mother's belly into the womb). So pre-verbal memory was acknowledged.

Current neuropsychology includes aspects of both approaches. It is known that emotion influences what memories are stored and what can be retrieved. And the brain is seen as registering, storing, and retrieving information. The information itself can be transmitted via any of the usual five senses – touch, smell, taste, vision, hearing – and is stored in areas of the brain which process information from these senses. For example, auditory-verbal information is stored in the areas (mainly in the left temporal lobe) which mediate speech comprehension and expression. If you have a left temporal brain injury (and are right handed) you may have deficits in both verbal fluency and verbal memory. If you have a left parietal-occipital brain injury you may find it difficult both to recognise and name things or places you have seen and to recall information about them. The hippocampus and the medial temporal lobe surrounding it mediate both new learning of information registered in any of the senses and topographical orientation. (Both functions involve the 'indexing' of information.) The frontal lobes and limbic system mediate both the determination of what is worth remembering and the search of memory for specific information. The cerebellum mediates both memory of movement and the execution of movement. Attention and arousal, mediated by sub-cortical structures, enable both registration and retrieval of information.

Memory is thus a 'distributed system'. It exists all over the brain. But so does thought. And as the evidence accumulates that brain damage in any particular area affects both the cognitive functions mediated in that area and whatever memory functions depend on the area, the distinction between memory and thought is not clear. The father of neuropsychology, Alexander Luria, paid very little attention to memory as such in testing people with brain damage.

Apart from being distributed throughout the brain in specific areas, memory is also a general brain function. Immediately after a head injury, even a light concussion or 'ding', memory for anything and everything may be impaired while the person may continue in activity 'on automatic'. And to complicate matters further, in complex brain functions an injury in a particular area may disable a wider function. Luria compared this to a tune being played on the piano: a single defective piano key means that the tune cannot be played completely, but this does not mean that the tune, or even the missing part of it is 'located' in the defective piano key. Functionally, Luria's tune (although he did not use the vocabulary of systems theory) is 'distributed' among the keys of the piano.

Memory may also be distributed in the body as a whole, as when we injure an ankle and tend to 'favour it' as if it remembers, or at least contains the history of what has happened to it. But whatever is true of memory turns out to be true of thinking – or to use the technical term (in the usual switch from old English to Latin), 'cognition'. One practical way of running a neuropsychological clinic is to call it a 'Memory Clinic'. (I have worked in three, for over twenty years). It sounds less frightening to the patient than a 'Brain Injury Clinic' or a 'Dementia Clinic'. But in practice, memory problems are almost never found without other cognitive problems. The exceptions are 'amnesic syndromes' where the hippocampus and medial temporal lobes have been damaged by brain fever due to toxins such as a huge excess of alcohol, or malnutrition, or infection. In amnesic syndromes mainly new learning takes a hit. But careful examination does show signs of other less notable deficits. And cognitive deficits are almost never found without memory problems.

Being something of a minimalist all around, I find memory an artificial construct in the context of neuropsychology. I cannot separate it usefully from cognition (thought). And since I grew up in Northern Ireland, where Elizabethan habits of speech survive, I have been aware since childhood that 'mind' and memory do not need to be distinguished. Shakespeare used 'mind' in the sense of 'memory', as most English speakers do when we say 'Mind your manners', meaning 'remember your manners.' When I was a child my school-mates would say 'I mind when you broke your ankle' – or whatever. Not too long ago I heard one of my Northern Irish nieces say, 'I mind when I was wee.' ('I remember when I was small'). In the Ulster-Scots still spoken by several hundred thousand people the word 'remember' simply does not exist. It is always 'mind'.

Memory as mind

It is a useful exercise to think of memory as 'mind'. Neuropsychologically, this makes sense. And it brings to the foreground the idea that memory is not, as usually supposed, merely to do with the past. It is *not* simply a record of the past – the engraving on a blank slate. Insofar as it refers to the past, then all our thinking does. Immediate memory is tested by the repetition of strings of digits. It is not quite in the present: it was, a few seconds ago. Similary with our thoughts. We recognise a friend. 'That's Bill!' But we are saying this after the event. There is a built in delay of 'a split second' (several hundred milliseconds if Cytowic is right) between the instant we see someone and the instant we are aware of seeing them. Various experiments show that we actually react to seeing Bill, in terms of autonomic responses of instant pleasure or anxiety, before we are conscious of seeing him. Thinking is as much in the past tense as memory. We cannot think or remember exactly in the present. So why distinguish between mind and memory?

The Spanish surrealist film maker Juan Buñuel, interviewed at about the age of 80 when he was developing dementia, said 'Loss of memory is loss of self.' He was, I suppose, thinking of remote memory – his own history crumbling away. In some dementias, such as Alzheimer's, remote memory remains intact for a while and the person has no such insight as Buñuel. But they are losing their minds as their new learning is eaten away. They cheerfully ask the same question again and again. They too are losing their selves, in a different way. Memory = Mind = Self. And clearly memory / mind / self is structured in the brain and body. Brain trauma or damage destroys it selectively, according to what part of the distributed system is eliminated by cell death. Primary progressive dementia such as Alzheimer's disease marches through the brain according to determined paths eliminating memory, mind, self. Neuropsychologists from Luria onwards have been disposed to be monists: they cannot separate mind from brain or body. The usual dualist options are unconvincing. For example the idea that the brain is just like a radio being acted on by an outside consciousness (the soul), so that if part of the radio breaks the consciousness cannot make itself heard. Or the Popper and Eccles idea (brilliantly refuted by McGinn) that consciousness 'emerges' from the brain. These cannot be accepted easily in the face of so much evidence that the mind / memory / self is so obviously structured in.

If electrodes are applied to parts of the brain (as originally by Wilder Penfold), they may trigger actions (unless the person is chemically restrained), or emotions, or memories. At least the streams of images that are reported are assumed to be memories. We do form the present from the past.

Images of the future

If J W Dunne is right, and the experimenters who have replicated his methods, and individuals like myself who have followed his procedures, then dreams contain an amount of information about the future. This may not be so neat as Dunne's assertion that they contain one third of each – past, present, future. In fact past and present must from a neuropsychological perspective be put together: any present image is composed of past images. But from the cases I cite in Chapter 6, the images of the burning store and of the balloons, I accept that some dream images 'come from' the future. Why is this not true of the images in memory?

Dreams access the mind (= memory) when the usual executive control functions (ECF) are asleep. The neuropsychological evidence is precisely that the frontal lobes which govern ECF are inactivated during sleep. The lid is off. (The lid being the necessity for purposeful activity when awake). So the dreamworld – perhaps as Panksepp proposes the world of other animals, with smaller frontal lobes proportionally to the rest of the brain than humans, in their waking hours – is awake. (The Australian Aboriginals have an idea that before the world we presently live in, we lived in a dreamworld). Most of our dreams contain a jumble of images that we recognise, but with new elements. We may, to take a banal but frequent example, find ourselves making love to a person we have never seen but who is nevertheless present in the dream in great detail. Or we may find ourselves flying. Robert Graves, of all people, proposed in a book on dreams that perhaps we were dreaming of an evolutionary future when we *shall* fly. But more likely the sensation is the familiar one of swimming, except that we are doing it in air. It can be argued that everything in our dreams is a combination of what we have already lived.

In this view if I dream of hurrying to miss a boat and having to stop to look for a lost ticket, then all the images are previously known. But this is *not* true. The ticket may be of a type I've never seen before. The face of someone in the dream may equally be new. They are only composites in the sense that anything we see in our waking life can be broken down into composite elements.

There may be language in a dream but it overwhelmingly consists of *imagery*. It may be jumbled and 'timeless'. But, as in the hurrying dream (and these are very common) *time* is expressed in imagery – for example the boat pulling away from shore. Perhaps birds dream of not having their nest ready 'on time' – not obviously through images of clocks and thoughts that 'soon it will be May', but as images of uncompleted nests along with sensations of eggs about to be laid. Perhaps birds have precognitive nightmares which turn out to be true. We do.

What about waking precognition? There are states of reverie between dream and awakeness. The kind of state in which Coleridge wrote Kubla Khan, perhaps, or the brief 'hynopompic' or 'hynogogic' states we experience between sleeping and waking or vice versa. But leaving these aside, why do we not experience thoughts and images while awake that predict the future? Presumably because we are too intent on living the present or even contemplating the past. When we contemplate the future, perhaps in fantasies of what we want to happen, we are not convinced that the fantasies will come to pass exactly as they come to mind. Otherwise we might frequently have experiences such as imagining someone then seeing them on the Tube or bus, and take such experiences for granted. Perhaps this is what *déja vu* is. Dunne supposed that this was simply the recognition of something one had seen in a dream. Which makes a kind of sense, as we know we forget our dreams very quickly. So when we see a one-legged banjo player at the entrance to the Tube station and have a sense of *déjà vu,* we may be able to accept the idea we have already dreamed this. In fact many people who have not read Dunne and know nothing of the theory of dream precognition have the *sense* that they may have dreamed this image before. They do not have the sense that they have *thought* it before. Our thoughts and memories simply do not churn up random images to that extent. If I am sitting writing a clinical report and my mind fills with random images I know I am falling asleep, I am beginning to dream. It seems that waking we are too occupied with living, with the necessity for constant orientation and re-orientation to the present, to allow dreams into our minds. Our frontal lobes are too active.

Damasio has pointed out that even autobiographical memory is 'an aggregate of dispositional records of who we have been physically and of who we have usually been behaviourally, *along with records of who we plan to be in the future.*' [My Italics.] And I suppose that if we are capable of memory of the future then memory will inevitably contain records of who we *have been.* Memory looks perilously like a continuum in which past and future are mixed. It is timeless.

In dream we find ourselves looking forwards through certain images. But we don't know it. If waking memory / mind worked the same way as dreams, then a delirious or confused person after a bang on the head or during a brain infection would come up with images and thoughts from the future as well as the past. This does not seem to happen. Falstaff on his death bed babbled famously about green fields. People who have lived through near death experiences describe their life flashing before their eyes – that is, their *past* life. But of course it can be argued that they don't have a future one. Could it be that a dying person experiences intensely his or her past life precisely because there is no more future?

Future memory
What in neuropsychology we call 'future memory' or 'prospective memory' is simply intention. 'I must water the plants this evening'. Then I do it. We do not recognise this narrowly defined 'future memory' as memory of the future – just as memory of thinking of the future. But if memory is

mind, perhaps all thinking of the future is a kind of future memory. Of course it is not very accurate. I may think I'll go to the theatre tonight but then have to cancel it because I get flu. But then past memory is not very accurate either. It is greatly influenced by our dispositions (what we want to be true), as in the 'false memory syndrome', and by the process analysed by Elizabeth Loftus in which every time we remember something it becomes changed slightly, so that eventually our memories are memories of memories... This may be because, as Damasio says, 'When we recall an object, we recall not just sensory characteristics of the object but the past reactions of the organism [ourselves] to that object.' Our dispositions affect our past memories, so why not our future ones?

It might be useful to do studies of the material that emerges in a delirium, or when a person is babbling away with a fluent dysphasia. In all that 'word salad' are there images of the future, as in dreams? What about the word salad of some forms of schizoprenia? Or the more simple statements such as 'I am dead' from someone who ends up dead soon afterwards by his or her own hand? Or, how many of our fantasies do, in fact, come true? How many of our passing 'flashes' of words or imagery about this or that refer to what will happen some day? These questions have not been explored except on the 'fringe' of science. J W Dunne found that if he took a novel before reading it, and paused to let his mind go blank (not as easy as it sounds), he would often 'flash' on an image which would turn out to be an important one in the novel when he read it. 'One had merely to arrest all obvious thinking of the past, and the future would become apparent in disconnected flashes.'

Drowning sailors who are were revived, in the old days, told stories of how they had been to 'Fiddler's Green' – a village where fiddlers played around the green. Released by imminent death from their past and from the need to orient themselves (never more impossible, I imagine, than when drowning in the timeless, spaceless sea), are they seeing their future? After all, they survived.

I cannot even say the jury is out on such questions. They have not been properly argued in court. If they were, a court might be more receptive than a university faculty or scientific congress. I was once instructed by a solicitor to examine a man who had been claiming for 14 years that he was one Joe Jones (I am changing the names in this account), with a certain birthdate, and obtaining social benefits under that name. He had a psychiatric history and, oddly, he would have been eligible for the benefits if he had applied for them under his own name. Nevertheless he was being prosecuted for benefit fraud. His defense was that he actually thought he was Joe Jones with the specific birthdate. Unfortunately there was a real Joe Jones with the same birthdate who understandably resented it when the police turned up and accused him of benefit fraud since he was in fact a good citizen with a regular job. Hence the prosecution. When the solicitor proposed that I should examine the defendant to see if he had any memory deficits, I laughed down the phone. A psychiatrist had examined him and said he was suffering from 'psychogenic amnesia' but actually to take an identity and believe, for 14 years, that you *were* that person seemed to me too far out. (Most psychogenic amnesias last for a few months at most). I took on the case because a neuropsychologist colleague had already examined the defendant for the prosecution and a judge had ruled that he needed a fair deal and should be examined for the defense. When I met the man it turned out that his memory functioning was intact for everyday matters, but his autobiographical memory up to 14 years ago was a blank. He nevertheless believed he was Joe Jones and his date of birth was such and such – even though he had been told a few years ago that he was in fact Bill Brown. He had even gone to check out Bill Brown's background and met a woman who said she was his mother – but he did not believe her. I am omitting much of the detail of the story. The main point is that although such a story does occasionally happen, and is attributed to psychogenic amnesia where a person under great stress leaves his home and arrives somewhere else and assumes a new identity, normally the story can be set right. In this case it could not be. The Court was faced with either a huge lie or the longest case of psychogenic amnesia on record. The London newspapers got something out of it: 'Experts disagree on stolen identity.' Under cross examination my colleague put the accepted academic view very clearly. I put the view that there is always a first time and that in this case the amnesia might be setting a record but was nevertheless possible. And this is where the difference between a university professor and a Judge comes in. The Judge was quite willing to accept (within reason, and there were various arguments) that this was a one-off case, where the professor would almost certainly have said it was 'impossibly rare.' In other words the judge did not have to consider probability theory and could accept the singularity of the case, so long as he knew, as my colleague had to agree, that it was not *im*possible. He ruled not on scientific but legal (in effect ethical) grounds, that for most of the 14 years the defendant had genuinely believed he was Joe Jones, but that once he had been alerted to the possibility he was Bill Brown and that someone else was Joe Jones he should have stopped claiming the benefits under the name of Joe Jones. He gave 'the Defendant, otherwise known as Joe Jones' a conditional discharge and everyone was happy – even my colleague on the other side.

The relevance of this to 'future memory' is that the Judge, and the rest of us in the court, including the Defendant (according to what he said, and he was a convincing man: I speak as someone who has examined dozens of psychopaths in cases ranging from murder to fraud) found it puzzling how he could have latched onto the name and birthdate of someone he had never met. I think it is possible that in the crisis that drove him from home (the typical trigger for psychogenic amnesia) he 'remembered' that he was Joe Jones with a certain birthdate – as for about ten years he turned out to be.

Yes this is far-fetched, but if a few dozen such cases were unearthed and carefully investigated, perhaps more could be learned about future memory.

In his memoir *The White Goddess: An Encounter* Simon Gough describes his role when aged 17 in a complex love story in the early 1960s involving his great-uncle Robert Graves and others including Graves's then 'Muse'. When the dust settled, Graves's wife Beryl explained to Gough:

> Everything that's happened has been a self-fulfilling prophecy… By which I mean that no one could have done anything out of their own free will, try as they might… The vision – and it *was a vision* – was Robert's, and everyone has played their part in it as he foresaw they would.

Events in Graves's life had played out a scenario he had described 15 years earlier in *The White Goddess*, 'a grammar of poetic myth.' The ancient myth as he had described it had been manifested among 20th century people. Had he made events happen? Hardly: there were several strong willed people involved in the drama, and they were not acting parts he had detailed to them. Could it be that he had foreseen the drama and mythologised it 15 years earlier? Was this future memory?

Certainly Graves did not believe in time. 'There is no time', he told me in Oxford the year after the drama Gough describes. And some years later he wrote to his new Muse:

> I am the expression of a combination of genes, which I chose myself. Reincarnation... would mean a different set of genes and inherited memories. If I am then asked, 'Then at what time did you take these decisions about yourself?' I answer 'Time is a convenience, not a fixed irreversible flow, which man is capable of disregarding (in the sense of fixed fated occurrence)... From the point of distant stars, I will not be born for millions of light-years; but I have my fixed place in the universe and this can never be altered. If therefore I am asked, 'When did you decide on your birth?' I answer: 'In the moment of death, when alone I have a full conspectus of my life, which is a sort of capsule containing an endless circle, head awallowing tail.'

Crazy stuff! Must be dementia (he was in his seventies)... But Julian Barbour in *The End of Time*, 30 years later, proposes that why we *experience* time in a timeless universe is that the universe contains 'time capsules', defined as 'any fixed pattern that creates or encodes the appearance of motion, change or history.' And 'the mature brain is a time capsule. History resides in its structure.' (And I know from Barbour that he did not know of Graves's theory of life as 'a sort of capsule', which in any case was in a private letter. That Graves's best friend at school was Barbour's father is just another 'more-than-coincidence'.).

In a timeless universe why should memory not travel both ways, forwards as well as backwards?

The outflow of perception

Research on vision (see R L Gregory) demonstrates that the simple 'inflow theory', originating with Helmholtz, that all sensory input acts on a more or less 'blank slate' in eyes and brain, is invalid. Instead there is evidence for an 'outflow' theory, that eyes and brain select from the chaos of outside impressions, and that the selection partly determines what is seen. Our consciousness is the view or perception of the world that our biology and experience predispose us to remember.

Just as an organism is surrounded by fields, it also, even if it is as small as a single cell (as Gurwitz's experiments with the mitogenic radiation of cells suggest) may have its own field. All fields may interact, at least locally. If they do so universally there must still be room for some sort of local selection process. This may well be in the form of a *resistance* against being totally overwhelmed by *all* impressions. This opens the way to a suggestion that the selection and focusing processes of consciousness involve a resistance to what is *not* selected. In the longer term the brain's forgetting of unnecessary information is accompanied by cell death. In childhood and adolescence the brain cells we are born with are 'pruned' by a selective dying off called apoptosis. A similar Darwinian survival of the fittest theory has even been proposed for memories.

To return to the idea of an 'output' theory of perception rather than an 'input' theory, I find this equally unsatisfying. Again a purely mechanistic approach to living processes becomes caught in an either / or scheme, and cannot adapt to a simultaneously two way process. There is no room in this purely mechanistic model for a process in which when I look at a cow, the cow itself is also in some way participatory. This idea may seem very far fetched, and I do not go so far as to think it would apply if I were looking at a stone. But again it is consistent with quantum physics where the observer interacts with the observed, and, before that, with the ideas of Ernst Mach. When, in a conversation with the physicist Julian Barbour about Popper's 'objective knowledge', I set out a pitch for 'inter-subjective knowledge', using the analogy of several people at different points outside a fence comparing notes on their separate views of a cow in a field, Barbour remarked: 'The full picture of the reality would have to include the cow's view of you.'

Mystical explanations of perception are equally unsatisfactory, and tend to involve ideas of us being 'receivers' tuned to certain sensory wavelengths. Tuned perhaps by God. Or more subtly by an accumulation of millions of sensory impressions from which an apparently purposeful selective power develops. Even Popper and Eccles' book on *The Self and its Brain* becomes mystical (under the Eccles influence) in proposing a rickety theory of 'emergent consciousness.'

Where is consciousness?

It is normally assumed that whatever goes on inside our brain was once outside. Imagery, for example, is of things we have seen or (as in a nightmare, or one of those reconstructed but imperfect memories that, according to Elizabeth Loftus, are the norm) a fusion of various sub-images we have at one time seen. 'Imagination' is thus a special kind of memory. Believing is seeing. At least believing, on the basis of previously registered external experience, is internal seeing. To take a simple example: a six month old baby crying and looking toward the door presumably has some image in its brain of mother entering the door, as she has done before. A newborn baby may similarly cry but presumably has no visual image of mother: it is normally assumed to cry only from animal appetite or internal pressure, although it must have a memory of sensual contact with mother. It already recognises her voice, for example, which it has heard in utero. A child several years old may also cry for its mother, but it is also capable of 'self-conscious' reflection such as 'she doesn't like me any more'. To rearrange these three levels of consciousness along Damasio's lines: the newborn is aware of internal needs (proto-conscious); the six month old is conscious of remembered sensory contact and imagery (core consciousness); the older child is self-aware and socially aware of the wider situation in which it has a place (extended consciousness). It can be argued that all these levels of consciousness are indistinguishable from memory. (Awareness of an internal need depends on the remembered previous occurrences of these needs – otherwise it would be merely a reflex, like a yell when someone treads on your foot.)

This seems to be a simple picture, and if it were sufficient it would not be relevant in this book. But there are two rather startling anomalies. First, instances of precognition (and perhaps of psychokinesis, although this is more far-fetched) suggest that what 'should' always first occur externally sometimes occurs first internally. Second, instances of information and imagery somehow 'at large' in a 'field' suggest that what 'should' occur internally sometimes occurs first externally.

When we encounter another person or a situation we are conscious of many sensations, perceptions and thoughts. But there also many levels of unconscious perception. The simplest of such perceptions have been demonstrated in neurophysiological experiments in which information is given to, registered, and stored by one part of the brain quite unconsciously while another part is paying conscious attention to another source of information. This is probably only the tip of an iceberg. We know a lot more than we are aware of. In fact this is an acknowledged phenomenon in 'psychogenic amnesia'. When 'the Defendant aka Joe Jones' in the legal case described earlier was informed about his early history he said that, once he had got over his disbelief via other people's explanations, he *knew* the woman claiming to be his mother was his mother but did not *remember* she was his mother.

Many of our actions are taken in the light of perceptions and information of which we are not even aware. A crude but powerful example is the fact that we are sexually attracted to other people by their smell – which we do not consciously register – but which may, for example, remind us of one of our parents. Panksepp has done experiments with adult male rats, long separated from their mothers, where he smears a little of their mother's breast milk on the vagina of one of many female rats in a group, then introduces the male rat to the group. Guess which female rat it instantly mates with…

In creative thinking, normal awareness broadens or intensifies to take in areas of perception which are normally out of awareness. The experience of writing a poem, for example – that is, if the poem is 'inspired' and not simply concocted – is 'mind expanding.' Graves described it as 'thinking on several levels at once'. Different levels of language meaning burst into awareness simultaneously and in a rhythm of their own – rather like a sudden sound of polyphonic music. Bach wrote on each of his manuscripts 'Gott Sei Dank' (Thanks be to God). Mozart listened to symphonies sounding in his brain while he played billiards with friends and drank white wine – then went home to the tedious and wrist- cramping business of writing out the score.

At such moments – and perhaps even for the rat and his mate, as much as for human lovers or for any of us in an inspired moment of work – the usual boundaries between our internal world and the external world seem to become totally permeable. Or perhaps to be perceiving in both directions. As I once found myself writing in a poem: 'and us / merely the walls of consciousness between two living flows.'

Pulse-waves are contained within the pulsation of every organism and every cell. It seems that although pulse-waves contain a regular beat, unequal phase pulsation is only found within a membrane. Thus, although pulse-waves may exist without pulsation, pulsation always contains pulse-waves. Is this true of thought processes? It is intuitively assumed that associative thought, dominated by a stream of images, is more 'primitive' than cause and effect logic dominated by the antithetical weighing of evidence. But as Levi-Strauss has shown, even the most 'primitive' tribespeople are almost compulsive reasoners in cause and effect terms. (Even if more 'civilised' people would not accept such premises that if someone looks at you long and hard he is casting a spell on you. But on second thoughts, perhaps they would. A man I assessed who was slowly recovering from a brain injury remarked that he had punched another man in the face 'because he looked at me.')

Nevertheless the open image stream of associative thought is comparable to pulse-waves, and the closed reasoning of logical thought comparable to pulsation. Where then is the parallel to the fact that pulse-waves occur *outside* the pulsating organism, for example in the atmosphere rippling and sparkling on a sunny day?

It is generally assumed that all thought or mind is contained within the brain. But there is no adequate physiological evidence for this. Popper (no logical slouch, after all) agreed with his more mystical co-author Eccles that 'mind' acts in and through the brain, but it is not identical to the brain. Damasio and Panksepp can offer more convincing arguments for the sense of self occurring as the most conscious part of the brain at any given moment (it usually includes areas of the pre-frontal cortex) registers the activity of less conscious areas. But they admit that a full explanation is still to be developed. And on the other hand, there are occasional disturbing experiences in the life of some people which suggest that consciousness may not be localised in the brain, or even in the body. The most cited example is the 'out of body' sensation of some people who survive a 'near death' experience. This sense of hovering above one's own body looking down on it in a bright light (usually) has recently been linked, through experiment on patients undergoing operations, to activation of a region of the angular gyrus, just beneath the pre-frontal cortex. But activation of a given brain area is not proof that this area originates the experience: it may simply 'mediate' it.

Where are the imagery and language of free flowing thought located? It may be accepted that logical thought, traditionally empirical, matter-bound, and concerned with checking imaginative conjecture, is in some way a development of physical bodily function, and based on physical trial and error testing: its functioning even suggests a membrane, with logical checking of insights being the imposition of a barrier against an otherwise unrestricted flow. What if thought and memory are mediated (for want of a better word) in the brain, but not always contained in it? Is there any evidence that the image stream, like pulse-waves, can exist *outside* the organism?

If by some extraordinary possibility there is a field of external consciousness which interacts with internal individualised consciousness, this amounts to a *duality* of consciousness, only rarely perhaps (as in a poem?) fused into one.

Past lives
In the 'Mumbo Jumbo' 1970s, when Krippner and Ullman were applying a more rigorous methodology than Dunne's to the study of dreams, there was more openness about the 'paranormal' in mainstream science than there is now. Quite respectable psychologists, for example, were interested in the question of whether the mind is in fact anchored in the brain, or whether it has or is an external 'field.'

In 1974 I attended a psychotherapy training workshop in Boston given by a psychiatrist, George Dillenger, from California, who was an ardent mystic. As an adjunct to the workshop, participants were invited to send, some weeks beforehand, to California, $30 and their name and date of birth. This minimal information would be given to a psychic, called Greg Tiffin, who would make a short tape containing a 'past life reading', which would then be transmitted and interpreted by the psychiatrist – who was of course, so the implication went, qualified to pick up the pieces if someone fell apart. I do not believe in reincarnation, but I sent the money and information out of curiosity, and because psychologist friends (it was, after all, the 1970s) had recommended the workshop.

When it was my turn to 'work' with Dillenger, he told me that I had had only one 'past life', that of an over-sensitive English boy who had been a soldier in the First World War, had not been able to adjust to the harshness of the army, and had died on active service, but of illness.

He provided an interpretation of this. It showed that I had been incapable of survival in the world in my previous life. Now my mission in my present life was to prove: 'I can.'

I had rather hoped that my past life would prove to be that of Sir Walter Ralegh, one of my favourite poets and men of action. But no such luck. Instead, I had been an over-sensitive failure. I almost cried. The 'past life' certainly made an impact on me: *I recognised it as a disowned part of myself.* For indeed, it *was* a 'past life'. It first brought me back to when I was a small boy. The first two and a half years of my life were spent in the Sussex countryside, adored by my mother, my grandmother, and my kindly grandfather. It was a paradise of cowslips in the fields and walks along the lanes. But for part of the time the paradise was invaded by an external terror: there was an anti-aircraft battery just past the bottom of our garden, which fired on and off all night, and by night and day German V1 pilotless bombs were being shot down by fighter planes to fall in the local fields instead of on London, in massive explosions. I was well protected in this: although I became oversensitive to noise and bangs, and ran a temperature, my grandfather would stand holding me in the safest part of the old house whenever a V 1's motor cut off meaning it was going to fall and explode. I cannot remember this time, though I can often sense it in my feeling for the place I was born. One of my first memories is of my father, in an army uniform. He came back from the war (he had fought at Alamein and across North Africa) a physical and nervous wreck, and a disciplinarian (although after some years he got back his considerable emotional balance). We moved to Northern Ireland when I was three and at first stayed with my father's family. Plucked out of my Sussex paradise, I can remember the grisly fascination of my father's family's house: a panelled hallway, the walls covered with photographs of First World War battlefields strewn with corpses, and with swords and pistols and muskets, and photographs of my various uncles in army uniforms: militarism and loyalty come easily to Haldanes. I absorbed this into myself. As I grew up I dressed in my father's cast-off army equipment, was enthusiastic about shooting with a bow and arrow, and joined the cadet corps at school where I found I was a good shot with a rifle. My favourite books were by G.A.Henty, about various soldier heroes of the past. When as an adolescent I read Graves's *Goodbye to All That*, about the First World War, it spoke to both sides of me, the sensitive (the poet – though I am not sure poets like Ralegh were all that sensitive) and the 'soldier'.

In other words I, as a person, have always had two sides (associated with my responses to my mother and my father in early years). The sensitive side I have often supressed. At the time of the psychotherapy workshop I was quite thoroughly supressing it, wanting to make an impression as a psychologist, never mentioning my obsession with poetry, down-playing my European-ness in North America, emphasising my more brutal side in intellectual aggression. My 'past life' was not a 'repressed' but a *suppressed* previous existence – in *this* life, not a 'previous incarnation.'

So this explained everything!

Except: how could a psychic in California, armed only with my name and birth-date (no address, no place of birth), come up with so accurate a vision? I saw other people in the workshop being equally moved by their 'past lives'. In each case I thought I could see the trick: these were powerful but disowned parts of themselves that the psychic had been able somehow to zero in on. He had then dressed them up as previous incarnations (or perhaps he thought they really were) and added some simple moral tags and advice plus the mystical pedagogy of the 'task in this life is to learn'. And the psychiatrist, a serious although perhaps misguided man, was doing a reasonable job of helping certain people come to terms with these visions.

(One reincarnationist counter argument to the idea that past lives are simply early parts of present life is that the person has 'chosen' a life which is similar but in which this time around the lessons of the past life can be put into practice. This is, of course, unarguable because circular. There is a neo-Calvinism in modern versions of reincarnationism: once the free choice to be born in such and such a guise has been made, the script unrolls in a predestined and unavoidable way. In the 1970s I once heard it argued seriously that even a child burned to death by napalm in Vietnam had 'chosen' this fate in order to learn a lesson – as sadistic a pedagogy as any in Calvinism.)

As an experiment I later asked a slight acquaintance, a woman from one of the Southern States where I had never been, to give me a few names and birthdates of friends of hers – none of whom I could conceivably know. As she gave out each name and birth-date, I let my mind go blank and then seized on the *first image* that the information evoked. This is the procedure J W Dunne used when he allowed his mind to 'flash' on the contents of a novel he was about to read. In one case I 'flashed' on a priest, in black, standing with a bible in his hand, in front of a church tower . The name turned out to be that of a man who had undergone a personal struggle about whether to take holy orders, but had decided against it, and *put the idea behind him*. Other examples were less dramatic, but similar. And where I have known a person who has had a 'past life reading' which has moved them (some readings are unmoving and obviously 'off'), the same thing has occurred. An example is a woman who was physically massive, clumsy and sexually extremely inhibited: she was informed by a psychic in a consultation by mail (they had not met) that in a previous incarnation she had been 'a temple prostitute in ancient Babylon'. This is a fairly common 'past life' given out for women, perhaps sometimes manipulatively. But it fitted: her eyes were velvety, doe-like and warm; she had been as a small girl the special favourite of her father but since her mother had clamped down on what had seemed too sensual a contact, she had repressed her sensuality and developed a massive 'armour.'

What explanatory theory can cover this? The data are there. The reader can experiment with concentrating on a particular name of a person (previously unknown and provided by someone else) then letting the mind go blank and paying attention to the first 'flash' of imagery or words.

It seems that information is 'picked up' by the psychic once co-ordinates (name, birthdate) are given. But this is somewhat mechanistic – as if there is an information bank which the psychic can 'access.' Two things strike me: first the fact that the information is contained in a vivid *image* which is accurate in itself, and to which the psychic usually adds some reasoned fiction or analysis (he has to make a living); second the fact that the image is of a *suppressed* and *disowned* part of the personality.

It is as if the image available to this kind of psychic poaching is either easily detached from the person, or already detached. This is something more than simple mind reading, or extra-sensorily reading what another is thinking about. In the case of my workshop the psychic was 3,000 miles away. This kind of phenomenon is what tempts some theoreticians to throw out the concept of *space*. Perhaps the information travels in a flash. After all, quantum physics – whether or not scientific psychology wants to take any notice of it – does admit *action at a distance*. One physicist, Henry Stapp, has done experiments which he interprets as meaning that the laws of quantum physics allow even *past* events to be changed as a result of action in the present. But this has to be taken with a pinch of salt as Stapp's reading of quantum theory is an idealistic 'Berkleyan' one in which the whole universe depends on our ideas of it.

In a mystical interpretation we are all hooked into a psychic network, all facets of the same universal consciousness. But how strange if this network or super-consciousness or whatever it is, is populated with *disowned fragments* of personality. These are not simply some kind of effluent or detritus, thrown out into a pool, since they retain their power to move the person. Perhaps they are like ghosts, or past stages of (this) life, and can be explained by some holographic model, for example Karl Pribram's, of the universe. But here again idealism (ultimately solipsism in which matter does not exist) looms. What *material* explanation is there of 'psychic poaching'?

Some mystics and spiritualists apparently believe that a person is more susceptible to toxic or evil influences if he or she is not well integrated. This makes psychological sense. But the spiritualists' point is that the vulnerable person's aura or 'etheric double' has holes in it or is torn and fragmented. The data from UAP (unidentified atmospheric phenomena) sightings shows that the people most liable to claim contact with 'spacemen' have schizophrenia or have had previous psychic or 'out of body' experiences (which damns them for researchers), whereas the people who merely observe the phenomenon without becoming 'contactees' fall within statistically normal parameters in terms of health and occupation. Again this makes psychological sense. But if each organism exists within the electromagnetic field of its mind, then fragmentation, or lack of integration, might occur also in the field.

Fancifully: perhaps when a person disowns part of him or herself it is thrown out into the periphery of the person's field. By 'it' I mean the imagery concerned with the disowned part – the memory which has been suppressed. Perhaps Galton's 'cellar' is a far reach of the person's life field. But the memory has the specific quality of being ejected from the person's *emotional* experience. When it is brought back into consciousness, through simple recall, it then evokes emotion. When imagery is 'out of sight, out of mind' it is also out of emotional range: emotions are felt physically, bodily. The experience of my current self is emotionally close to me. I do not believe a psychic could poach from my life field the imagery of my love for my wife or children, for example, because this is part of my everyday emotional life. My father used to lament that he never dreamed of my mother when in North Africa during the war because he was thinking of her every day.

That psychics pick up information in the form of imagery (or of words or numbers as it were flashed on the screen of their consciousness) is suggested from the experimental work of the Rhines and their successors. It is also consistent with the neuropsychology of memory – almost always stored as visual imagery, like dreams. The psychics do not seem to pick up logical or linguistic thought sequences. Perhaps these are too closely bound to a person's experience of him or herself as an enclosed pulsating unit.

7

THE ORIGINS OF POEMS

Inspiration

So many modern poets, when interviewed in literary magazines or writing in prose about their own work, admit to an intention or a plan in writing poems about certain themes, it is fair to assume that many 'poets' write poems as a deliberate exercise – as a painter chooses a theme and paints it. And poetry competitions bring in hundreds of so-called poems written to specifications – e.g. 'a poem not more than 40 lines long on the subject of Manchester.' Many poets are not reluctant to see poetry as an art which can be studied and learned. Universities and colleges offer creative writing classes and workshops in which poems are concocted. 'People can't put on an opera but they can write a poem' according to a 'future poet Laureate' Simon Armitage, in 1999 appointed 'Poet of the Millenium' and producing to order 'a bitter, excoriating work, lambasting celebrity culture, the tackiness of striving to conceive a millennium baby and Anglo-American foreign policy.' More recently (2006) he has written an 878 line poem on 9/11 ('That is my son dying' etc., although Armitage is not known to have been near New York on 9 / 11. He admits to writing the poem based on TV reporting). He has said in an interview, 'I always wanted to be a poet.' But this is a give away. If you are a poet you don't need to want to be one: you just hope you'll write poems – something out of your control. A poem is not as Armitage says 'built'. As the German poet Wilhelm Lehmann wrote (I translate): 'Poems must be worked on. But first they must *originate*.' Lehmann believed that when words and rhymes came together in a poem *things* were coming together in nature. 'Rhymes occur when things are allowed to meet up together, not word chimes.'

The current English poet laureate Carol Ann Duffy has remarked: 'The idea of great poetic genius is a thing of the past. We are more democratic now.' I do not like the idea of 'great poetic genius' either. Some of the most powerful poems have been anonymous ballads or one off occurrences in the lives of people who would never promote themselves as poets but have briefly lived the experience of being one. (One of the greatest poems in Irish, for example, is Eileen O'Leary's 18th century Lament for her husband Art. Another great 18th century poem in Irish English is the anonymous thieves' ballad, 'The Night Before Larry was Stretched.') But it is false democracy to insist that everybody is a poet. Poems are rare.

A 'real' (as distinct from fake) poem, or a 'durable' (as distinct from dud) poem is radically different from any deliberately concocted verse in two ways (at least): its very reality or durability, and its origin in what used to be called 'inspiration'. But poets who have written real and durable poems, when they discuss their origin (sometimes they keep quiet about it) almost always make it clear that for them the experience of writing poems is in itself unusual, quite apart from the poem itself.

Of course anyone can claim to have been inspired. The children's writer, Enid Blyton, for instance gushed that all her stories 'wrote themselves', as did the trash novelist Barbara Cartland who could turn out a book every few weeks in just as long a time as it took to dictate it. But some poets have described their experience more succinctly.

A E Housman described how on long walks

> As I went along, thinking of nothing in particular, only looking around me and following the progress of the seasons, there would flow into my mind, with sudden and unaccountable emotion, sometimes a line or two of verse, sometimes a whole stanza at once, accompanied, not preceded, by a vague notion of the poem which they were destined to form part of. Then there would usually be a lull of an hour or so, then perhaps the spring would bubble up again.

Robert Graves described the

> poetic trance, which happens no more predictably than a migraine or an epileptic fit... All poems, it seems, grow from a small verbal nucleus gradually assuming individual rhythm and verse form.

Robert Frost wrote that

> For myself the originality need be no more than the freshness of a poem run in the way I have described: from delight to wisdom. The figure is the same as for love. Like a piece of ice on a hot stove the poem must ride on its own melting. A poem may be worked over once it is in being, but it may not be worried into being. Its most precious quality will remain its having run itself and carried away the poet with it.

Even the reticent Thomas Hardy described his poems as 'impelled.' The even more reticent (outside his poems) Sorley Maclean admitted that he had written several of the metrically very intricate (in Scottish Gaelic) poems of his cycle Dáin do Eimhir (Poems to Emer) in one go, having woken up in the middle of the night.

For myself, poetry comes from a voice in my head – not someone else's (I'm not psychotic), but my own voice which is out of my control.

In this chapter I shall use some of my own poems as examples in looking briefly at what poetry *is*, as a phenomenon. (What poetry has to say about time will be discussed later.)

Poem 1, The Dagd

One morning at the end of March 2000 I was finding it hard to get up, as at 7.00 it felt like 6.00 because of the change to summer time. I really had to force my eyelids open. I joked to my wife that I felt like the 'Dagd', the Irish sun god with only one eye which was so big it had to be propped open. Then in the shower a poem came into my head:

> An ancient giant, in a farm cart dragged
> By labourers, his one remaining eye's huge lid
> Propped open with a spear: the DAGD.
>
> For the Greeks, Apollo's winged chariot scorched across the skies.
> But the Irish cut the DAGD down to size.
> Well, they would.
>
> Why should a sun god have two eyes?

I scribbled this down when dressing and at breakfast remarked to my wife that I had written a poem in the shower about the Dagd and how the Irish had cut their sun god down to size. 'They would', she remarked. I found myself thinking that in ancient Ireland there was also a sun goddess (Áine), who had probably preceded the Dagd in the tradition, and that the 'two Is' of my wife and myself could at times be 'one I' in the affinity of our thoughts.

The poem is true to the myth. It is not even a very personal poem in that it did not express a current conflict for me. My feeling when writing it was one of pleasure and amusement. I suppose I was aware of the Irish 'knocking' tendency in myself. But the timing of the poem is of interest: it occurs just after the change to summer time with its implicit theme of the sun's progress across the sky, and it is followed by my wife's echo of its key phrase, 'they would', and my thought about 'two Is' and 'one I'.

The poem can even be seen as the key part of a series. But although it is in the middle of the series, it is 'the last word' about the events it describes. I can add all sorts of thoughts to it – about the function of sun gods, about my fatigue that spring, about the affinity of 'two Is', and so on. I do not need to add the poem. Anyone can read things in it from their own lives. But it is still the last word.

Poem 2, This and That

When I lived in Prince Edward Island and worked in a Mental Health Clinic I was seeing my last client of the morning and listening to her account of a long-standing conflict with her husband when my mind wandered briefly to having made love with my wife the night before. A poem pushed its way into my mind:

> This needs that
> But this this,
> This that – more
> Than this or
> That – must be
> You and me,
> This in that

Come to this –
O that this!

But for what
Does that that
That is you
Need this this?
That round this
Come to that –
☉ this that!

Those were the words, as they came. But as often happens, they came with a rush of other images. In fact the poem turns out to contain no explicit imagery at all: *this* and *that* stand for all sorts of things which simultaneously flooded my mind. Meanwhile I was half listening to the client droning on and I was faced with the problem of retaining the words in my mind. (I forget verse easily, whether my own or other people's.) I kept repeating them in my mind like a spell as I went through the motions of discussing the client's dilemma for another ten minutes or so, then showed her to the door and ran to my desk to write out the poem. On reflection I think my state of consciousness throughout was 'trance consciousness', and in my interaction with the client I was 'on automatic', as when driving a car and talking intensely to a companion. With the client, I am afraid to say, my discussion was strictly speaking unconscious. (I certainly could not remember a word of it afterwards.)

If there was a trigger to the poem I suppose it was the contrast between the misery of the client's account of distance from her husband and my sense of closeness, that morning at least, to my wife. There was *that* (last night) and *this* (now). There was most explicitly the *this* and *that* of the penis and vagina – or the *that* and *this* as they became in a sense interchangeable. The poem is about *sex* in the sense of the 'cut' (the original meaning of the word, which is connected with *scis*sors) between the two sexes – *that* and *this*. *This* and *that* (and vice versa) are also the other differences between my wife and me, including our conflicts as a man and woman living together (like the conflicts of my client). The poem celebrates *vive la différence* between the sexes, but feels the difference in a sort of horror too.

As so often when writing a poem my emotion was simultaneously a deep horror (a sort of dread) and delight. Complicated!

I was so struck by how the bare bones of this poem, the *this* and *that*, carried so many meanings, that as an exercise I wrote out the whole meaning of the poem in prose. It took three pages, and of course as prose is rather tedious. The most difficult thing to explain about the experience of writing a poem is the density of the imagery, feeling, and thought that all rush into consciousness at once as it occurs (whether it is being written out or sounded in the head when no pen and paper are to hand). It has appeared out of nowhere – as of from the *un*conscious. Yet it does not feel as if it has all been thought through in the unconscious and prepared for writing out.

I suppose a theory of inspiration might be just that: unconscious processes work away and produce a poem (just as they can solve problems as we sleep) which then, when it is complete, rises to the surface. Some people say it feels as if a poem is 'dictated' to them. Yet, as Graves has said in discussing this idea, there is no doubt the voice is one's own, and the poem 'about' one's own experience. As Graves puts it, the poem has many 'layers of meaning' but these are all experienced at once during the writing. The poem is not planned, it is a revelation. Or as Graves puts it, 'a miraculous event in non-history.' It feels in that sense timeless – or at least out of time. It seems to include past, present and future. In the case of This and That it includes many images from the past – but this is true of any writing. We have only experienced the past and what we call the present is in fact an awareness of what has just happened – in that succession of 'nows' (literally a fraction of a second long) we are aware of. The sexual imagery of This and That is from last night, and the conflict imagery from a minute or so before in the client's account. The present in the poem is an intense experience of many layers of the past. But there also seems to be a future – in the feeling of mixed dread and excitement which accompanied the poem, as so many. A sort of cry from the heart: Is *this* what life is like? Is this what the future holds? Is this – *this and that* – all that we are? Are we personal or impersonal, are we Seán and Ghislaine (just names anyway), or are we just a man and a woman, or just *this* and *that*? If so, how awful! Or how wonderful!

So, again, the poem has the last word. It says more in a very short space than can be said about it in a very long space in prose. And it has a very odd relationship with what we call time. For one thing, in the poem *everything is happening at once* – or at least huge spaces of meaning are condensed into a small space. The parts and the whole are simultaneously visible. Huge areas of time are condensed into a shorter area. A poem may therefore seem timeless, or to change time. So that Shakespeare, for example, writes sonnets about how they defy and defeat time – along with other sonnets that give in despairingly to time – and others (the best perhaps) that both win and lose, at once, against time. Here we are back to the paradoxes of poetry we see in Parmenides.

The greatest paradox may be that while a poem may feel as if it is changing time or transcending it, it is in itself bound by the rhythms of time. This and That has an unusual form, with lines of three syllables and three stresses – RAT-TAT-TAT, RAT-TAT-TAT – but it cannot escape its own rhythm. It pulses – though not in the usual way.

Poem 3, Mumm's Champagne

> When you cracked open the Mumm's champagne
> To celebrate our first whole night –
> In the new sheets you had bought, striped blue and white –
> You didn't know how much pain
> Would follow from being together.
>
> But our love has not turned vinegar,
> Though it's no longer bubbly – a blood red wine
> Pressed from the last grapes of two ageing vines
> Whose limbs would have to be snapped
> To get them apart.
>
> We have grown around each other, and here
> Are the same sheets, the stripes faded and worn,
> And a bottle of something more modest than Mumm's.
> The champagne spurts
> And though we forget nothing, nothing hurts.

This is, I hope straightforward enough to tell its own story and need no explanation. But it originated in a loose series of coincidences. First, we had had guests and my wife was short of sheets so she dug out some old ones we had not used for some years – sheets we had first used many years before. We made the bed together, joking about the sheets. We were spending the evening together for the first time in a while without family or other company. I had bought a bottle of champagne – of a sort (Cava, really). After years of bringing up children we could no longer afford Mumms. We had been talking a lot, and rather painfully, about how she had given up her work to be a full time mother (a 'Mum' of course) for some years then had to retrain in a new profession. That week I had been ordering some crates of wine via the Internet and had received a lot of publicity about 'old vines.' When I wrote the poem the following day, it took its place among the events it describes. Perhaps it made amends for some of the pain: our relationship started in very difficult circumstances and the striped sheets were emblematic of it – alternating light and dark like our first days together, and much of our life together. (And my wife is very dark, while I am fair). The beginning of the poem also revealed something of the 'fate' of our coming together: the fact that the champagne she had bought was Mumms had no significance of 'mother' or mothering at the time: she was French-speaking and we spoke French. I had paid no attention to the brand name beyond its being a good one. But now it was a prophecy. The poem's last line also resolved a dilemma between us: it was possible not to forget painful things but nevertheless to be no longer pained by them. And of course the poem is about sex too.

There is nothing magic here, but again the poem is a sort of 'last word' on a situation – or the last event in a very long cluster of events.

Poem 4, Canadian War Memorial, Green Park

> I think of Bertram Warr,
> A leaf fallen from a plane
> In the year I was born –
> Another poet gone –
> And Isaac Rosenberg.

> If they were to meet
> (Wherever they might be)
> They might talk of Stepney:
> Life in a slum,
> A rat in a bombed house carrying a crumb.

Poems to be published should explain themselves and I am not sure this one does, as not every one will have heard of Warr and Rosenberg. But it illustrates a more simple sort of occurrence of a poem than This and That.

I went on my own on Remembrance Day 2000 to the Canadian War Memorial, in Green Park, London, at 11.00 a.m., and stood for a while thinking of my father who served in the war (and who lived some years in Canada as a young man) and of poets who died in the world wars. Warr was a Canadian who wrote a few lovely poems and was shot down over Germany when on a bombing mission in the RAF, in 1943, the year of my birth, when he was in his mid-twenties. Rosenberg, much better known as a poet and painter, was a London East End Jew who was killed, also in his mid-twenties, in the trenches in 1917. I was thinking about their common fate and the complete lack of connection between them. I was staring at the hundreds of bronze maple leaves under the trickling water of the memorial and the hundreds of real maple-like golden leaves strewn across the nearby grass, fallen from the plane trees in Green Park. On a wave of sadness the first three lines of the poem floated into my mind, with an awareness of the double meaning of 'plane'. Then as my mind moved to Rosenberg I suddenly realised there was a connection: Warr had written a poem about exploring a bombed out house in Stepney, where Rosenberg had grown up. The image of the rat carrying a crumb is from Warr's poem.

The poem formed itself out of these sudden connections. In a way the poem is just an intense form of thought in rhythm and rhyme. It contains no new happenings. I suppose I felt, for the moment, as if I was somehow connecting Warr and Rosenberg. But the poem makes several connections. There is obviously a difference between a poem and a 'series'. But they have in common the making of connections among things which are not otherwise connected. Again, thought also does this. But in writing a poem one feels part of the thought process and somehow not responsible for it all.

Take a list of the 'ingredients' in this poem:

> The Canadian War Memorial
> Green Park
> Bertram Warr
> 1943
> Isaac Rosenberg
> plane trees and fallen leaves
> bronze maple leaves
> a bombed out slum
> Stepney
> A rat carrying a crumb
> Myself.

I suppose it would be possible to concoct a poem from these ingredients, and the connections between the two 'war poets' and the war memorial are immediate enough. No doubt I am making these connections, from what I know. But writing the poem was involuntary and it felt as if Warr were somehow contributing to it with his poem about the rat.

Actually, there are more connections than I was aware of as the poem came to mind. I have only just recalled when writing this that one of Rosenberg's best known poems, Daybreak in the Trenches, is about observing, and identifying with, a rat. And for all I know the bombed house Warr explored was the former house of the Rosenbergs – flattened as it was by bombing long after they left it.

Even a simple poem like this has more connections than I could have consciously thought up – and may have more I have not thought of yet. As I think of new connections, they are like buried layers in the poem. For example, now I have remembered Daybreak in the Trenches and its 'insolent' rat looking at Rosenberg, this will be part of any re-reading of my own poem. I even find myself wondering if Warr – who could have read Rosenberg's poems – was looking for Rosenberg's house in Stepney and thinking of Rosenberg's rat.

There are also connections or resonances from the names of the two poets. 'Warr' is obvious. But Rosenberg evokes 'rose' and the most popular sentimental song in 1916 and 1917 was Frederick Weatherby's Roses of Picardy. Rosenberg was killed in Picardy.

Poetry and the brain

I should add that as a neuropsychologist I haven't a clue about the origins of poetry in the brain (if they are in the brain, which I sometimes doubt). No convincing experiments on inspiration could be done, I think, because it is a rare and unpredictable phenomenon. There are all sorts of psychological essays and books about 'creativity' but the net is cast too wide so as to include hundreds of so-called 'artists'. A poem is not planned. (If it is, then it is not a poem but a damp squib – like the 'occasional' verses of the former poet laureate Andrew Motion which are a cause merely for parody and laughter). But it may be highly organised even in its first draft. My guess is that in inspiration, when everything happens at once, all areas of the brain – including the emotional systems, intellectual processing systems, executive control systems, and memory systems – are simultaneously aroused. This arousal is by defnition autonomic.

The classicist Housman, who had no pretensions to science, described meticulously the autonomic reactions he felt when writing (or even reading or thinking of) a poem: his whiskers bristling, hair standing slightly on end, cold chivers going down the spine, a feeling in the stomach that he describes by quoting Keats's remark that the memory of his girlfriend Fanny Brawne 'goes through me like a spear.'

The neuroscientist Jaak Panksepp, who has identified seven emotional operating systems of the brain (SEEKING, ANGER, FEAR, PANIC, LUST, CARE and PLAY) has speculated about the biological origins of the 'chills' and 'thrills' sometimes felt when a person is moved by music. His description echoes Housman's (which he did not know at the time of writing but found accurate when they were shown to him):

> We even love to hear sad songs – especially bittersweet songs of unrequited love and loss. A common physical experience that people report when listening to such moving music, especially melancholy songs of lost love and longing, as

well as patriotic pride from music that commemorates lost warriors, is a shiver up and down the spine, which often spreads down to the arms and legs, and indeed, all over the body.... An intriguing possibility is that a major component of the poignant feelings that accompany sad music are sounds that may acoustically resemble separation DVs [distress vocalisations] – the primal cry of being lost or in despair.

Possibly systems involving opioids or thermoregulatory systems are involved. 'Thus when we hear the sound of someone who is lost, especially if it is our child, we also feel cold. This may be nature's way of promoting reunion. In other words, the experience of separation establishes an internal feeling of thermoregulatory discomfort that can be alleviated by the warmth of reunion.'

Panksepp concludes that 'The study of music will have profound consequences for understanding the psychology and neurobiology of human emotions.'

Panksepp's own work is already revolutionising neuroscience with its emphasis on emotion, and he is collaborating on studies of whether birdsong is changed by playing music to the birds. But as Wyndham Lewis wrote in *Time and Western Man*: 'In music the sounds *say* nothing.' Poetry is so dense with meaning compared with music that it may be the closest approach to a description of reality available to us. I have always resisted the (possibly profitable) idea of writing about the psychology of poetry, because poetry is so much deeper and denser in meaning than psychology. To write about the psychology of poetry would be like writing about planet earth solely in terms of cartography.

The pre-existence of poems

A poem feels like the final word. But it also seems to pre-exist. When writing, it feels as if the poem is already there. There is a sense of excitement at revealing it, and often (for me) a sense of horror – 'Oh No!' But one does not always grasp it clearly in the first draft. So one works to 'get it right' (a phrase Robert Graves used). This getting it right may require patience in the area of finding the right word: usually the right words are there from the beginning, but where a word feels wrong it seldom works to go to a dictionary or thesaurus. If one waits, the word will eventually turn up. It is like what happens when one cannot remember something. I often tell Memory Clinic patients not to *try* to remember a word 'on the tip of the tongue' (known as the 'TOT' phenomenon). If they wait, it will float to the surface sooner or later (unless they have Alzheimer's disease). In the case of a poem, this feels like the last word to the last word. But other things to 'get right' are rhythm, sound variations, and sometimes rhymes. Although again, a rhyming dictionary does not help and feels somehow wrong. The poem must, on all its levels, be left to complete itself as it will usually do spontaneously if the poet comes back to it a few times and re-reads it attentively. The poem has a life of its own. Sculptors apparently sometimes feel that they are chipping away at the block of material to find the sculpture which is already in there.

Although a poem consists of words, it is not just a flow of words and sound. The words have exact *meaning*. And the poem has a *form* like a sculpture. It looks like a poem on the page, and its sounds and rhythms have form. In fact the most common poetic metre in English, the five stress line – as in 'Shall I compare thee to a summer's day?' – has a form that matches the biological rhythms of heartbeat and breathing. The five 'beats' to a line correspond roughly to usual ratio of five heartbeats to a breath. The line is roughly the same length as a breath – and it often has a rise two beats in (Shall I com*pare* thee to a summer's day?) that corresponds to the pulsatory shift from shorter inbreath to longer outbreath. In the poem its shape (the lines on the page) and its sound correspond.

A well known poetic form in most European languages is the sonnet. It has fourteen lines (which makes its length correspond roughly with a minute) and one or other complex rhyme scheme. Some poets have reported the experience of writing a poem straight out and arriving at the the last couple of lines or the end before realising it is a sonnet, complex rhyme scheme and all. (This has happened to me once or twice, e.g. in Persistence, quoted later). No doubt the Elizabethan sonneteers who churned out hundreds of them could write them in their sleep, and obviously the knowledge of these poetic forms is stored in the brains of those who read them. But it still feels uncanny to write in a form before knowing it is that form. Again the poem pre-exists.

Of course much modern poetry rejects form and is in 'free verse', but as Robert Frost said this is like playing tennis without a net. Most free verse is dead on the page. Being entirely rooted in normal speech it is a left hemisphere language phenomenon, with no input from the 'musical' right hemisphere. It is not poetry in the sense this has been understood for millennia. It is merely prose pretending to be poetry. As the American poet Trumbull Stickney said, poetry is musical thought.

Free verse satisfied a rebellious prejudice against form – not unnatural after centuries in which hack versifiers killed form by regularising it too much, as in the 18th century 'heroic couplet' in which identical rhythms succeed each other, much like plastic imitations of a sculpture coming out of an assembly line. It also coincided with what Wyndham Lewis called the 20th century 'time cult' in which under the influence of Bergsonian and Jamesian philosophy, life was seen as a flow of unconnected sensations, without form or meaning. Even the painting of the period reflects this – from Impressionism, though Expressionism, to Abstract painting. (When I go to the Quai d'Orsay gallery in Paris I don't go up to the top floor where the Monets are, I stay downstairs with the Manets and Courbets. I like *line*, in painting as in poetry. And this is part of my character – my personal meaning and form. Although the poem once written is 'out there' as if it has pre-existed and demanded to be written, I cannot deny it has come from my own experience.)

Poem 5, Black Hill

> Along the moorside, scattering sheep,
> Clambering over walls of black stone,
> Under the lark's twitter in the sky sphere
> And the hobby hawk's begging wheep,
> I've climbed. I'd thought I was alone.
> But *they* were there. They *are* here
> On the ridge in a round barrow
> Tussocked with grass, crumbling down,
> Boulders tumbled into its crown.

I lie on the grass and press my ear
Against a boulder. Dimly I hear:

Who I am you don't want to know.
How it is you don't want to know.

Dimly I see blue eyes scrunched narrow,
White cheeks and forehead, yellow hair.

You don't want to know who I am.
I was killed with me Dad and me Mam.
Death is more than you want to know.
What it's like you don't want to know.

Yellow-haired girl, you don't want to know
How all is changed, nothing is changed.
I lie despairing on this slope.
In this world I find no hope.
Here I am with the whiffling air,
And wheeping hawks, and larks so high
I can't tell where they are in the sky.
No flowers for a grave, the moor is bare.

Draw me a circle on the stone
With your finger, a cross in the circle – so.
And put your lips to the circle – so.
This is my forehead, kiss my forehead
So I can feel that I'm not dead.

I've kissed the cross in the circle – so.
Along the moorside, scattering sheep,
I descend to the hazy valley below.
Three hares start from my track as I go.

> Yellow-haired girl asleep in the ground,
> Under the grass and stone of your mound –
> I don't want to know. You don't want to know.

In 1998 I gave several organisational psychology 'workshops' with colleagues at the Moorside Hotel in Derbyshire, for managers and executives of a large industrial firm. The workshops were routine although the company was good. I had never been in Derbyshire but had some interest in it because the landscape looked like Ireland in some ways and my maternal grandfather, who was born in Yorkshire, had the surname Riley (no connection to the Irish Reilly: it means 'rye-lea', a rye clearing) and this is most common in Derbyshire. So I imagined some kinship with the place. In May the days were getting long and although on one day it snowed there was usually warm sunshine so some of us would go for walks in the early evening, before a late dinner. We usually walked across the road and up a winding road to the West. From this the hill which loomed above the hotel to the East was more visible, crossed by stone walls and dotted with sheep, and I suppose about 500 feet high from the road. One evening I decided to climb it on my own, clambering over the drystone walls and slogging up to the top, a long ridge with another valley down to the East. At the higher end of the ridge were several collapsed Bronze Age or Iron Age tumuli, just as so often in Ireland on hill tops. I had the thought '*They* are here', and after that I walked around in a sort of trance with lines of the poem forming in my head. As I did what I did the lines formed. As I descended the hill afterwards with the sun setting behind me the poem completed itself in my head and I wrote it down before dinner.

After writing the poem I was curious about the name of the hill but no-one in the hotel (where few if any of the employees were local) knew it. When I got home I looked it up on a map. Black Hill. So that gave me a name for the poem, and very appropriate too, given my mood. I was next up at Moorside in June and I climbed the hill again one evening, this time in a spirit of curiosity. I stood studying the chaotic heap of boulders where I had lain down and listened through the stone. I noticed some shreds of paper and went to pick them up. They were fragments of a torn up letter and seemed to have been stuffed under one of the stones. At first I thought someone had used them to wipe themselves, but the stains on them were from bird droppings. The handwriting was in black ballpoint and in a childish hand, smudged as the paper had obviously been wet by rains. It had obviously been a love letter: there were words about disappointment and love and being let down. And on one of the fragments I could discern clearly the word *Sean* (without the accent – i.e. not Seán). I could not tell if the letter was written to Sean or from Sean. It is not a common name in Derbyshire but it exists there as everywhere since Sean Connery made the James Bond movies. I felt a shiver down my spine. Was it a letter from the girl with yellow hair?, I thought confusedly. I brought the letter home with me, thinking I would piece it together. But I never did. I don't read other people's letters. Perhaps it is best not to go too far into the origins of poems.

Where a poem comes from
A poet does not know how many poems he or she has 'in him' or 'in her.' As Martin Seymour-Smith has written, the worry about whether there are going to be any more poems is diagnostic of the true poet. (The fake poet, of which there are many, is impatient with this and simply concocts fake poems.) Although it is impossible to ascertain what happens in in the brain during the writing of a poem, perhaps what Julian Jaynes calls the bicameral mind is at work, and the right hemisphere is speaking to the left. This theory of Jaynes' was long though to be merely eccentric but recent evidence from neuroscience 'split brain' experiments supports it. As Louis Cozolino puts it, the evidence suggests 'another "will" residing in the right hemisphere.' Most recently, Elkhonon Goldberg, a former student of Luria, in *The Executive Brain* has spelled out a theory and findings that suggest a new definition of hemispheric function. The 'visual' right hemisphere is also (as is now well known) partly verbal but its role is to process new information and encode it in more general visual images and verbal schemes, before passing it on to the left hemisphere for more precise verbal encoding in the person's history. (This echoes previous theories of the brain dealing with 'fluid' and 'crystallised' information.) Goldberg discusses Jaynes's theory, long supposedly refuted by evidence that the 'visual' right hemisphere cannot 'speak' to the left, and re-habilitates it – in a neat example of how a Popperian 'refutation' falsifies a theory but new evidence from a new perspective suggests it is true.

Jaynes cites evidence from cuneiform and other inscriptions in the Eastern Mediterranean 'cradle of civilisation' in the second millennium BC that when the gods spoke to mortals it was always in verse – usually in a dactylic metre (- ^^ / stress, unstress, unstress) the utterances of the oracles and sybils (usually un-trained countrywomen who opened themselves to the words of the god) in the first millennium BC were in the same metre. And amazingly, it also breaks out in the 20th century trance utterances of Afro-Brazilian 'Umbanda' mediums.

I do not completely rule out the possibility that the yellow haired girl under the mound on Black Hill actually spoke to me from the past. But in that case she would have somehow been an English speaker and therefore buried there in recent centuries. It is more likely that anyone buried under that mound would be a speaker of a Celtic or pre-Celtic language, in which case the voice I heard may have been a voice of my own but somehow speaking 'for her' or 'as her'. (Which does not mean I 'made her up', or that she is an unconscious part of me. I felt her as an external presence.) But more usually for me the poem I write has a voice – my own – as if 'I' am speaking quietly and urgently to 'me'. I hear the poem and obey it in writing it down. It takes me over. There is nothing else in mind. This reduction of all awareness to one source is typical of hypnotic and other trances. It is trance consciousness. It cannot be invoked at will. I only find myself writing a poem or fragment of one five to ten times a year or so if I am lucky. In theory a SPECT scan of my brain might reveal, during the writing of the poem, the 'lighting up' of my musical right temporal lobe as my linguistic left temporal lobe lights up also – or in Goldberg's scheme the passing of fluid information from the right to be crystallised in the left. Perhaps the lobes excite each other into 'musical thought' with a powerful autonomic / emotional impulse via the sub-cortex which mediates rhythm. But I cannot live my life under a SPECT scanner waiting for the miracle of a poem. (I imagine Simon Armitage would oblige if he were asked to write a poem under a scanner, but it would not be a poem.)

The phenomenon of inspiration can be distinguished from possession, where the entranced oracle or prophet is taken over by a voice not his or her own, and does not remember afterwards what the voice has said. Another of Plato's gifts to posterity was the idea that poets are mad when inspired, possessed, and 'out of their senses.' Jaynes quotes the Iliad to refute this slander: 'Say to me, Muses… for you are goddesses and *and are present and know all.*' We have seen how Parmenides in *Peri Physeos* listens to the voice of the un-named goddess for almost all the poem, after an initial 'kindling' description of the young man, himself, being rushed into her presence. Parmenides' poem is a kind of dialogue, in that the goddess addresses him directly and anticipates his objections to her uttered truth. It is a highly complex poem, full of reflective consciousness, but it reads as if written in a trance and its intensity is such that it may have been written over one session – not impossible even if it was originally somewhat longer than than the 161 lines that survive. I have myself written a 100 line poem (Ingratitude to the General – admittedly hardly on Parmenides' level) while 'taking notes' in a management meeting. And one of my more intense poems, Desire in Belfast, 106 lines long, forced itself on me in two chunks in the same morning, while I was driving, so that I had to pull onto the verge to write them down.

Personally, I am never more in my senses than when writing a poem. But a shift in cultural consciousness, at least, has taken place since Homer. In writing a poem I realise that the voice is not the Muse's but my own – while nevertheless having a superstition or devout sense that the poem would not occur if the Muse was not present, in the form of the woman I have in mind when writing the poem. Most of my poems, even when not love poems, are written 'to' a woman. (The Elizabethans would have said 'through' a woman.) And for most women poets I think there is an equivalent presence of an addressed man.

At any rate the voice in my poems is me. So it makes sense to suppose that it is one part of my mind (or brain, or field of consciousness) speaking in verse to another.

As I have noted earlier, I personally lack what Jaynes calls the 'analogue I' – the images in mind of the self seen by the self. This is partly temperamental. As I suspect it is in Jaynes who is so sure of the omnipresence of the analogue I that I think he must have been something of a narcissist. But partly, in my case, it is a choice. I remember deciding, shortly after I had written my first poems, and under the influence of a girlfriend who was unusually spontaneous and naïve, that I would reject all narcissistic images – as if seeing myself in an internal mirror – and live only through my own feelings and eyes. Martin Seymour-Smith described inspired poets as necessarily naïve, as distinct from self-consciously 'sentimental'. (This distinction originates with the German 18th century poet Schiller). Probably naivete is necessary in a poet. Thomas Hardy for example was obviously naïve. But he was also cunning as a fox.

Wherever the voice of a poem is coming from, and even if it is describing events in time, it seems to come from 'outside time'. This too works in brain terms. In my doctoral thesis I suggested, on the rudimentary evidence available in 1977, that the right hemisphere was faster than the left. Goldberg now confirms this.

The poem gathers together events, images and feelings from different times, into one. Its 'last word' is timeless. The poem comes from nowhere into somewhere. The 'nowhere' may be the generalising right hemisphere and the 'somewhere' the particularising left hemisphere. The left hemisphere quantifies time, makes it linear, measures it. The right hemisphere must be involved in spatialising it. Time cannot be said to be invented by one lobe or the other, it must be invented by both. I suspect that in a poem, as in a dream, the frontal lobes (on both sides) although they retain their organising function are somewhat de-activated in their inhibitory function. The frontal lobes are known to be the main areas of the brain involved in making time estimates and judgements: they are critical and inhibitory, and it is clear that if the poet is going to open his or her mind to the poem coming from nowhere, critical hesitation must be set aside. (It is reinstated when revising the poem and eliminating lines or words where inspiration has failed. As Laura Riding said: 'criticism is death'.) The driving rhythm of the poem, presumably mediated by sub-cortical structures such as the basal ganglia, but deriving from the pulsation of the whole body, reminds us that timelessness and time coexist. We cannot escape time.

More-than-coincidence is a sudden sense of *meaning* in a recognized cluster of events. The meaning may be at first sight trivial – a link among bald-headed shop owners and fire, or among the vocable 'iv' in certain names – but it is recognised with emotion, as non causal links often seem to be. We are freed from our usual cause-effect chains. More-than-coincidence transcends time.

Similarly a poem. It may turn out to have drawn words and images from widely scattered moments in time, but together (linked as in an a-causal event cluster) they now have meaning and they transcend time – even as the poem that embodies them is alive with the rythms and counter-rhythms (pulsation and pulse-waves) which make us aware of time. Perhaps the poem moves us with a sense simultaneously of mortality and immortality, extinction and eternity, time and timelessness.

8

PERSISTENCE IN THE COSMIC OCEAN

Seeing Time
We see time. Obvious examples are an ECG or EEG where the electrical activity of heart or brain is graphed in time, the trace of a respirometer recording the breathing cycle, or the pulsation of the blood under fMRI. But we also see time if we stand and watch waves pounding on a beach, or the pulsing and coupling streams of an aurora. And insofar as time exists in the Shape Universe it must be structured into the shape. I find myself imagining it as corrugations, like those left by the tide on the sand of a seashore. This is consistent with the ideas that function precedes form, or that form is 'frozen history' – except that the history is happening *now*. The tide comes in and goes out and there are ripples on the sand. But in a Shape Universe the ripples *are* time. So are the traces of the ECG or EEG or respirometer. They are records, if seen from a certain perspective – when the tide appears to have gone out, or when the ECG trace is examined in the clinic. But while they are forming – as the tide is crossing the sand, as the ECG trace is printing out and the heart is beating under the electrodes – they are not records but time itself. Time in action, one might say, if one thinks in terms of the Time Universe. The shape of time, in terms of the Shape Universe. Faced with ripples on the sand, or sea waves, or ECG traces, we confront the ancient questions of Heraclitus and Parmenides, and the relevance of Su Shi's equally ancient conclusion (worth quoting again) that 'If you think of it from the point of view of changing, then Heaven and Earth have never been able to stay as they are even for the blink of an eye. But if you think of it from the point of view of not changing, then neither the self nor other things ever come to an end.'

We live in two worlds, two universes at once. What is more there are two observable patterns in time traces: the equal phases of the non-living (ripples on the sand) and the unequal phases of the living (the ECG).

Most of us, for most of the time (as it were: it is impossible to avoid time language) 'go with the flow' and accept what we are seeing as part of a transitory flux. As in the Shakespeare sonnet, 'Like as the waves make towards the pebbled shore, / So do our minutes hasten to their end.' But Shakespeare also described himself as 'at war with time.' We are part of time, but at the same time (!) we feel we are separate from it. We do things to it: we *spend* it, we *waste* it, we *lose* it, we *take* it, we *make* it, we *prolong* it, we *economise* it, we *manage* it, we *run out* of it, we *make up* for it, we *squander* it – and so on. I propose that our separation from time is due to the fact that although we are surrounded by equal phase pulse waves, we ourselves pulsate. And we *sense* our separation from the surrounding flux whose pulse waves impinge on the membrane within which we pulsate. We are discontinuous in the flux's continuum

In terms of the Shape Universe as described by Barbour, the membrane which separates us from time at large, the surface of our pulsating bodies, may correspond to the boundaries of a 'time capsule'. But Barbour does not discuss life as distinct from non-life, and I doubt if he would accept this analogy. In his theory even a fossil or a book are time capsules. But it is interesting how yet again language draws him into metaphors which imply 'time' meanings. Capsules in nature (excluding the artificial capsules manufactured to contain drugs) are always organic: the seed pods of seaweed, for example, or the cysts in which unicells go dormant.

We feel time inside ourselves, most notably as our heartbeat and breathing. We perceive it outside ourselves – through seeing it in auroras or ocean waves, hearing it as a drumbeat in the night, feeling it as a pulsing breeze on our skin. We may even smell it or taste it, as when night falls and drying seaweed on the rocks or the night-stock in a flowerbed send out their aromas according to the diurnal cycle. And we have all sorts of timers in the brain to integrate internal time and external time. I maintain that since our internal time is pulsation and external time is pulse- waves, the difference between the two constitutes our awareness (consciousness). I think this is true for any pulsating organism, such as a jellyfish, although as its much smaller neural network shows, it has much less to be aware of than we do. As Popper says with uncharacteristic humour, the amoeba cannot face its own hypothesis – it is part of it. Similarly with the jellyfish. We are obviously less part of the cosmic ocean, our environment, than a jellyfish. We are more separate, more distinct. And something more than in the nursery rhyme:

> A jellyfish swam in the tropical sea
> And said, 'there's no-one I'd rather be
> There's nothing above and nothing below
> That a jellyfish doesn't know.

All the same we are pulsating, aware organisms, with defined boundaries (even if these 'membranes' are simply the boundaries of force fields) in a universe which is either without boundaries or so vast as to be without them from our point of view. As far as we can see or our instruments can detect, this universe is saturated with pulse-waves. It is a cosmic ocean in which we swim or float. Unlike the earth's ocean which is held by gravity, and in which whatever stops floating or swimming falls to the bottom, there is no direction in the cosmic ocean. There is no absolute motion, only relative motion.

If the Shape Universe is the correct hypothesis, then the cosmic ocean is as it were frozen, a vast mass of ice, a plenum in which nothing moves but the sense of movement comes from our relations with all the other things, alive or not, which are frozen into the ice – within which we are all directly or indirectly interconnected.

If the Time Universe is correct, then the ocean is expanding and flowing outwards in all directions and very gradually we and all the ingredients of this universe are losing touch with each other and decaying (entropy), although perhaps here and there in the flood conditions permit certain structures we call life to become organised, to grow, and to interact (negative entropy).

The Multiverse, I suppose, implies that there are billions of parallel oceans but that they are, at least at certain moments, in communication with or coincident with each other. These moments are quantum *discontinuities* in the apparent continuum of the cosmic oceans (plural).

Whichever of the three universes turns out to be correct, the view of ourselves as pulsating organisms in an ocean of pulse-waves is a valid way of understanding how we perceive time. And since quantum discontinuities will not go away, even if the remaining classical physicists want them to, there is room in the cosmic ocean for *anomalies* to occur – time-fractures, precognition, poems, miracles.

Orientation

The first necessity for any organism in an ocean is to orient itself. A newborn baby looks around or squirms to reunite itself with its mother. A swarm of jellyfish or a flock of starlings visibly switch directions to orient themselves to their environment, most immediately in the form of currents or winds or the direction of light. All organisms have extensive orientation structures in their nervous systems. Birds even contain a sort of organic electromagnetic compass. In human beings, neurological damage shows itself in a loss of cognitive orientation to place, time and person, and of physical orientation in the form of balance and directed movement.

Orientation is individually structured. But it can also observably be a collective process. The school of fish or the flock of birds may dash and swoop 'as one'. It is assumed they all share a common perception, either of the environment or of each other, or both. But there are cases which have been carefully observed where a group of organisms act as one where they are unable to perceive each other and where each is in a different place. The phenomenon of thousands of fireflies in mangrove swamps, each isolated from the other, flashing on and off in perfect unison, was startling to early explorers and their accounts were not believed. These phenomena are known as *entrainment*.

If entrainment is seen as the simultaneous adaptation by organisms of the same species to changes in the environment – a sort of collective orientation – then it is not necessary to invoke 'paranormal' factors such as extra-sensory perception. The flock of starlings is no different from the crowd of thousands of spectators at a football match who 'as one' leap to their feet and roar when a goal is scored. It can also be argued that most of the thousands of spectators are thinking more or less the same thoughts during the football match, as they follow intently the progress of the game. A few might be thinking of other things. A few starlings may break off from the flock. But such procedures are dangerous. The stray starling, or the stray deer is vulnerable. When living in Quebec I often saw on the snow the tracks of small herds of deer, where one had wandered off from the rest and then its tracks were joined by those of a few wolves, and then eventually I would find the remains of the deer in a patch of blood-stained snow. What would be the risk of suddenly at a political rally shouting out 'Down with the Leader'? At English football grounds the authorities keep the fans of each team on separate sides of the stands.

Orientation, unlike entrainment, is individual. The animal or person adapts to the environment, using innate instincts (a word being rehabilitated in neuroscience as instincts can be mapped as innate emotional or appetitive systems) and learning from experience. This includes dream experience, as the working through of our waking experience in dreams includes orientation. I once dreamt I was looking at a map of an American city in which there were none of the green areas which indicate parks but there were green margins along some roads. I woke up thinking: 'There are no parks, but there are parkways.'

We all orientate ourselves consciously first thing on awakening and constantly throughout the day – sometimes with great pleasure. I remember my delight as a child learning the London Underground train map, with its different colours for each line, and its different lines connecting stations to each other at roughly equal distances. Even then I could see that the map did not correspond at all to a normal 'overground' map. The scales and distances were entirely different and much easier to grasp. I often now see children on the 'Tube', from all over the world, who are learning the map faster than their parents and who are guiding the family around. The map was designed according to a then unique principle in 1931 by an engineer draughtsman, Harry Beck, who had been laid off from his job with the Underground. It was at first rejected as 'too revolutionary' but eventually he won £10 for it. It was pointed out to him that it looked like an electrical circuit board but he said he had been thinking that for people underground the requirements of a surface map would not apply. His map is based on the *relations* and *connections* between and among stations.

The Tube map actually reflects a Machian universe where all that counts are observation and connections. Distances are irrelevant. And so is time: differently from a surface rail map, you cannot tell from the Tube map roughly how long it will take to travel from one station to another. Nor does this matter, as times / distances are short, and most journeys across London from one end to the other of an Underground line with 40 or so stations take about 1 ¼ hours.

In the 'many instants' of the Shape Universe the entire universe can be seen as orienting itself and re-orienting itself with each 'Now'. Each instant is a distinct set of relations among distinct entities. But we do impose time on our orientations within the cosmic ocean. We are individual points of awareness. (The word 'individual', meaning un-dividable, might as well be the Leibnizian 'monad' – a 'one'.) The universe itself consists of a 'set of sets' or a 'frame of frames'. At any given 'Now' we are oriented within a certain frame. For example we might be standing under the stars in the frame of our observation of the night sky. Only if we stand there a long time do we see any signs of movement of the stars in their slow revolution around the pole star. We may over a few minutes detect the moon's movement against the background of a constellation. From this perspective we impose time, in calling the rotation of the stars 'slow'. By contrast, if we look down a microscope into a drop of fluid which is swarming with unicells against a background of crystals and specks and streaks of micro-matter, this is a very busy scene. Cucumber-like paramecia come zooming across, vorticellae tumble across with their flagellae beating. Things happen 'fast'. But we can guess that from the vorticella's point of view things are less speeded up than for us. Or that from the point of view of a God looking at the solar system from a position relatively in eternity, the stars move fast enough. But we impose *our* time on what we see. And we agree with other individuals to observe the conventions of clock-time. Or at least we do in the industrialised world. Ask some North American native Indians who are used to living in their own good time to come for a job at nine o'clock and they may wander in separately all morning. Yet as the anthropologist Hugh Brody has recounted in *Maps and Dreams*, the same Indians who are 'incapable' of turning up to do a job on time will, when hunting for moose, split up in the forest at dawn, take their separate ways, and come together at various unspecified times during the day, as if by magic.

Our relationship *with* time is ambivalent. First we are surrounded by the pulse- waves of the force fields we live in. Second we are aware of the inner time of our pulsation. Third we learn to operate by social conventions of time and by clocks. *With* in old English has the sense of 'along with' but also the sense of 'against' – as in the phrase 'John fought with Jim.' Our sense of ourselves as individuals is bound up with a sense of being against time. The Indians described by Brody are most against time when they cannot bring themselves to turn up 'on time' for the non-Indian employer and are defining themselves most as individuals. Hunting together in the forest they separate without a word and come together on and off all day in perfect timing. Although separate in their awareness they are hunting 'as one'. To use Barbour's phrasing, their experience is not identical (as it would be in a football crowd or a riot) but it is shared. Not all agreement in thinking and behaviour is entrainment – which appears instinctual and mindless. Much agreement is 'common sense' among distinct individuals.

How does evolution fit in a timeless universe? Its very existence in such a universe is counter-intuitive. We know organisms and species evolve – literally 'unfold' – with time. But if we subtract time we are left with a series of 'nows' in which individual organisms *orient* themselves to their environment, and perhaps even create themselves from moment to moment (or here to here) in doing so.

The cause and effect model in which the environment is supposed through natural selection (Monod's 'chance and necessity') to have evolved an inner genetic programme in the organism to cope with the external world can be opposed by another more 'Leibnizian' model in which there is a correspondence between the organism's inner world and the outer world: they match in what Leibniz called the *Lebenswelt* (the 'life world'). In the early development of organisms their individual evolution (unfolding) occurs independently of their experience although there is a match between inner and outer worlds which are, as Spilsbury puts it, 'coupled.' Spilsbury updates Leibniz's question 'How can independent monads reflect or mirror the universe from a particular viewpoint?' to 'How are the cell-imprisoned genes environmentally oriented? Without access to the world, how can they "instruct" developing organisms on its ways? How is the innate worldliness of lower organisms possible? How has the eye been brought to light?' And, 'in some cases, the very objects of sensory attention and discrimination, along with the appropriate response, seem to be built into the sensory-motor system – which implies an amazing original coincidence of inner and outer.' Goethe, the poet, made a similar point in his theory of colours. Even the hard-boiled philosopher Colin McGinn makes it: 'Knowledge is a coincidence between the way the mind is intrinsically constituted, innately and otherwise, and an independent reality.' And in this coincidence, Spilsbury concludes, Mach-like or Barbour-like, that 'the places of things are all important.'

Continuum and Discontinuity

The Earth's ocean, like its atmosphere, is a continuum. There is a bottom and a surface but no sides. This must be the case in the surface of any sphere. So the odds are that the universe contains at least as many continuums as it does spheres (even allowing most spheres to be barren). But it appears the whole universe is a continuum. The 'empty space' I was taught about at school is now agreed by astronomers to be dense with fields of matter / energy and greatly more with invisible or dark matter than with visible matter. In this 'set of sets', there are oceans within oceans within oceans.

Quantum physics presents us with evidence of discontinuities at the micro level which two out of the three major physicists whose theories I have discussed (Deutsch and Barbour) and no doubt a mischievous part of the third (Hawking) agree must operate also at the macro level. I get out of my depth very quickly in discussing quantum physics because my mathematics are poor, and whenever the discussion becomes abstract I lose track. And so far we seem to be offered a choice of competing explanations for discontinuities. But there is no question they exist and that they turn the laws of classical physics upside down. In the current climate of stand-off among physicists it is encouraging to find Freeman Dyson (see Introduction) stating: 'I find it plausible that a world of mental phenomena should exist, too fluid and evanescent to be grasped with the cumbersome tools of science.'

A singularity, a one off event, is not of much use to science, since it can be neither replicated nor repeatedly observed. The series and precognition series and poems that I have described earlier are certainly not replicable – any more than you and I are, or any natural phenomenon except apparently for strings of genes (but even there 'peri-genetic' factors are being discovered that make the issue more complex than the reductionist Dawkinses of this world would have it). Not even two pine trees in a plantation are identical.

(Remember Leibnitz's point that identity is impossible). The trouble with replicable experiments is that they require that what Popper calls the 'truth target' be reduced to something quite small. So instead of researching the life of the elephant, we end up counting its droppings. This kind of bean-counting experimentation is 'normal' or 'drongo' science at its worst. In my own work I have to reckon with the random control trials (RCTs) of drug company scientists who cut so much variability out of their sample groups, to eliminate 'confounding factors' and make the groups 'squeaky clean', that their results have little relevance to real clinical problems, and furthermore often contradict each other. And the numbers – the mathematics – are an abstraction: even beans being counted are not identical.

So the possibility that a singularity can occur via a discontinuity is of as little interest to me as its theoretical opposite, the randomised control trial. I have left that sentence unedited as an example of how abstract statements require abstract language. But what I mean is 'So what?' Replicability is impossible. But systematic observations, or thinking carefully about unusual observations are not. I have at times deliberately written down all dreams I experience over a few mornings. At other times I have recorded in writing what strikes me as a series. Series stand out from the general background of events – the daily continuum. They seem to be following a different from normal pattern of connections – a discontinuity. As I wrote in a poem many years ago: 'Time and space are fractured by our affinity.' Emotion seems to have the power to disturb the continuum. Or, leaving aside cause-effect, emotions occur discontinuously as compared to everyday thought. In another poem which I decided to call 'Ambition', as I was not sure I would ever reach its goal, I describe my own persistence against the continuum. It can be renamed 'Persistence'.

> This hopeless swimming toward the source, bashed
> By waves that flush me toward a rocky shore,
> Plunging my head through green walls, lashed
> By strips of foam, I have no more time for.
> Time is on its side, sea through which I swim.
> I fight for me. I know the source is cold,
> The calm, relentless moon that tugs the rim
> Of ocean to a standing wave. More bold

To turn my back, bruised, wade in to the strand
Of some quiet cove, and make a life away
From hissing wave-tops on the unmoving land
Where gorse clings to cliffs un-splashed by spray,
And grasshoppers in tiny glades defy
The downward crush of gravity and sky.

The whole and the parts

We have a dual nervous system: the Central Nervous System (CNS) and the Autonomic Nervous System (ANS, sometimes known as the Peripheral Nervous System). Both connect our brains and the rest of our bodies. The ANS is itself dual. It has a 'sympathetic' component (most activated in conditions of anxiety or emergency) and a 'parasympathetic' component (most activated in dreamy or relaxed states). The ANS is the system of *INvoluntary* activity, and the CNS of *Voluntary* activity. The ANS is activated in making love (when excitation is mediated parasympathetically then orgasm is activated sympathetically). The CNS is involved in willing oneself to go to work on a sunny day. If we are unfortunate enough to be in prison (I am thinking of the prison of the Time universe), it is by the deliberate use of the CNS in controlling our behaviour that we get through. If we are lucky enough to fall in love, then the ANS has its day.

This duality in our nervous system presents a problem for the Shape Universe. It is all very well to say the brain is a time capsule, but if so it is a dual function capsule. No doubt the theory can accommodate this: Barbour is very clear that even in Platonia time exists via time capsules, it is part of that particular configuration structure. But the dual function allows us to live in two worlds: the time world and the shape world, the continuum of the ocean and the spontaneous lives of the organisms in it. The CNS allows us to direct thought voluntarily, and the ANS opens us to spontaneity. This does not mean the CNS thinks and the ANS feels. Feeling / emotion and thought are in constant interaction all through the brain. A poem seems to me to be very much an ANS event. It is not willed into existence, or if it is it dies the death and becomes mere verse. But it may be dense with thought – *feeling thought*. Robert Graves when he gave a talk at MIT in the 1960s met the mathematician Martin Deutsch (no relation to David so far as I know, but son of the psycho-analyst Helene Deutsch) who told him that what he, Graves, called poetic thought was *structural thought*. I wish I knew what he meant. (Both he and Graves are now dead so it cannot be checked). I wonder if it refers to the sense that the poem *is already there* – a structure that is revealed. It feels like something outside time – but it can be brought into time. Graves and Alan Hodge (in *The Long Weekend*, a social history) wrote: 'To use Army organisation terms – painting is properly the I branch, intelligent reporting; poetry is properly the O branch, active decision.' This does not mean the poem is itself action. But its 'last word' aspect means that it can change people's actions in the world of prose.

Perhaps the language of the continuum is prose, and that of discontinuity is poetry. The poet Laura Riding described poetry as 'an interruption in the life of habit.' The continuum, after all, exists not only 'out there' but 'in us'. Our duality is that we are pulsating individuals, sets in a larger set which includes pulse-waves. The fluid in which our brain floats is both pulsating and permeated by pulse-waves. Insofar as we are 'creatures of habit' we participate (we cannot avoid it) in the continuum. Insofar as we are open to interruption in our habits we have minds. Samuel Butler proposed that we have two kinds of memory, an individual memory, and a species memory which we experience as habit but which enables species to modify their goals and influence their own evolution. This annoyed Darwin no end. Darwin was invested in the Victorian version of the Time Universe, which he had more or less invented – Evolution – and every Tom Dick and Harry who opposed evolution came up with arguments that there was not enough *time* for evolution to have occurred by the mere chance of random selection. They were against time. Darwin was all for it. They were interested in morphology for its own sake and for its links across species *in the present*. Darwin was interested in it only for its record of the long history of evolution.

The Time Universe and the Shape Universe are two points of view (Su Shi again!), each structured into our brains. Sometimes one is up, sometimes the other. It is rather like Gestalt psychology (heavily influenced by Mach) where figure and ground keep switching: you look at a profile of a face in black and suddenly it becomes the partner of another face in white. It depends on how you look at it. Or strictly speaking, on orientation as you switch the focus of your eyes slightly each time, changing your point of view. Or in some cases, without moving your eyes as you 'will' yourself (presumably activating your frontal lobes to switch from one brain hemisphere to another via the corpus callosum) to switch perspective. Wolfgang Koehler, the proponent of Gestalt Psychology maintained that it was not perceptually possible to see the whole and the parts at the same time (i.e. instant), and it is not possible to see the two sides of a Gestalt switch at the same time either.

I find this too black and white. I have already mentioned and supported Goethe's view that the whole and the parts can be apprehended together, each making the other. In my main outdoor activity, field archery, there is much debate about whether a 'barebow' archer (one who does not use sights) should concentrate on the target or on the point of the arrow. The received wisdom is that only one or the other is possible, as in a Gestalt switch. But the great American longbow archer Howard Hill described his own practice as 'split vision': he focussed on the target but simultaneously considered the point of the arrow in relation to it. He proposed a simple exercise:

> Select any given object to represent the target to be hit, and focus your eye on the object. Using the right hand, closed except for the index finger, bring that finger into your field of vision.
>
> At first it will be difficult to keep from shifting your direct vision away from the original or primary object, but after some practice it will be easy to hold fast with your direct vision on the original object, looking at it primarily, while secondarily you will be able to point your finger at any other object inside the scope of your vision, without looking directly at either the finger or whatever secondary object you have selected. Keep both eyes open at all times.

Many archers, including myself, use this method. Most people see this way often in daily life – in observing a landscape, for example. Of course there is a tendency to focus on one thing at a time, but it is not always necessary.

I suspect the Gestalt 'switch' only works in those figure/ground drawings which Koehler and others used to illustrate it. And so an uneducated American bow-hunter refutes the Gestalt school of psychology...

Dualities
I am suggesting here a new duality. We have been hypnotized for centuries by the dualities first of body and soul, second of body and mind. Like almost all neuroscientists I am inclined to be a monist, not a dualist, when I study the functions of the human brain. I see no reason to consider body and mind as separate or that mind has somehow emerged from body. Although I agree with Boris Pasternak that 'philosophy is like pickles: a small taste of it once a day is enough', for a systematic discussion of the impossibility of 'emergent consciousness', I recommend the philosopher Colin McGinn.

However, *in our body* there is evidence of duality at many levels. There is the obvious duality of man and woman. But leaving that aside for the moment, there is a conspicuous duality in the CNS and ANS. The ANS mediates our pulsation, our spontaneity. It concentrates on the whole via the interaction of pulsation and pulse waves, living and non-living. The CNS mediates our interface with the pulse-wave universe by consciously resisting it or cooperating with it as necessary. It concentrates on the parts. Our pulsation is simply a different (unconscious) kind of resistance to the universal flux from the conscious resistance of our will. Both add up to what Kammerer called *Beharrung*, a sort of stubbornness best translated as *persistence*.

Awareness is consciousness as a function of perception. But there are clearly more complex levels of consciousness, perhaps as set out by Damasio. One unorthodox theory of consciousness is that of Julian Jaynes. As mentioned earlier, Jaynes is really discussing *self*-consciousness. His main theory, which in the light of evidence from anthropology and more recently from neuroscience is unlikely to be valid, is that consciousness is a function of language and that indeed language pre-exists it. He proposes that consciousness (in effect self-consciousness) did not exist until about 1,000 BC, and before that human activities were mainly unconscious but controlled by inner voices and hallucinations providing 'structions' (his own word) for unconscious actions. There is of course ample evidence that we can perform all sorts of actions unconsciously: we don't *have* to be conscious all the time. When humanity became conscious, Jaynes claims, we began to 'spatialize' our experience in our minds and to create mental constructs, including time. It is not necessary to buy into Jaynes's very peculiar main theory to find his discussions of what he calls the 'bicameral mind' stimulating. The bicameral mind is by definition a mind with two compartments – a dual mind. The compartments broadly involve the right and left brain hemispheres but Jaynes is careful to emphasise that this does not mean our sense of consciousness is located anywhere in particular: as he notes, it can subjectively be inside our own heads, behind us, or even on the ceiling. Although he does not discuss fields of consciousness he is describing them.

In the pre-conscious bicameral era, organising voices from one compartment (usually from the right hemisphere) tell the other compartment what to do. It still happens in poetry. One side of the duality is the telling what to do, the other the doing what is told. The most bizarre aspect of Jaynes's theory is that in this process the person is not conscious, but a sort of automaton. Jaynes states:

Now simply subtract… consciousness and you have what a bicameral man would be like. The world would happen to him and his action would be an inextricable part of that happening with no consciousness whatever.

This is reminiscent of Mae Wan Ho's statement that the action is the cause. And Jaynes's bicameral mind can be seen neatly as one of Barbour's time capsules: it just exists as a structure in a timeless universe which permits the illusion of time perception.

However, even Jaynes admits we are conscious *now*. And the evidence of modern neuroscience is in the opposite direction to his theory. It suggests that consciousness, at more complex levels than mere reactivity or even awareness, does not depend on language at all but does depend on *emotion* – which is common to both us and animals and has been present for millenia. Nevertheless the evidence also agrees with Jaynes's contention that 'We are… conscious less of the time than we think, because we cannot be conscious of when we are not conscious.' Many of our most complex actions, such as playing the piano, are almost entirely outside our consciousness: we may know we are playing but we do not know exactly what we are doing. We may even have a sense of 'flow'. The psychologist Mihaly Csikszentmihalyi has written a book on this state where we are so absorbed in an action that time is distorted and self-consciousness (what Jaynes calls simply consciousness) disappears. But this flow is partly an illusion. Like stream-of-consciousness it is a false perception of continuity. Like all our experience it is made up of successions of 'nows' – as when the eye looking across a landscape is composing its apparently stable image through tiny movements or 'saccades' which register a 'snapshot' on the retina every 10 milliseconds or so. And if flow itself is an illusion, all the more reason to dismiss the time-flow as one.

Throughout this book I have used the word 'flow' freely. It is a straightforward description of something we observe and experience. But there is a final irony in the fact that the flow is a construct of our perceptions. I have described the flow of pulse-waves as a continuum, and indeed we experience it as a continuum in the same way we experience water as a continuum as we swim in it. But it turns out that this continuum is a series of snapshots – what Barbour calls 'nows'. Out of the myriad snapshots taken by our vision we construct continuous colour and shape. Out of the myriad vibrations of a violin string or of the column of air in a flute we construct continuous music. Or for that matter we construct continuous sound out of the vibrations of a person's vocal cords. Or the archer's snapshots of arrow point and target merge into one.

Pulse waves in the cosmic ocean

Let us look at pulse-waves again. With their discrete ups and downs, their oscillations and vibrations (extending, we are told, even to the vibrating strings and superstrings that are the most basic matter / energy) they are the infinite number of 'nows' in the universe. We invent the continuum. We invent even the cosmic ocean. We are bigger in the universe than we think we are: or our ability to perceive constancy where none exists makes us large. Yes the universe appears monstrously huge in the way we are encouraged to see it in astronomy texts with their immense distances.

But perhaps what 'counts' in the universe is not the distance numbers but the numbers of relations between distinct entities. And among the entities of the universe we know of, we are the most complex – the most varied bundles of ingredients. It is a truism that without us the universe *as we know it* would not exist. But we forget this. Most of us can feel with David Stove's account of lying on his back looking up at the stars and being sure that 'there is nothing, however near or far, which takes or even could take the smallest interest in you.' But we also know that if we get up and look around and find our friends there are other entities who take an interest in us, as we do in them. And unless we are solipsists we believe that the universe exists whether or not we are interested in *it*. But we, the living (including the humble jellyfish or amoeba) seem to make a special intervention in the universe: we create, through our existence as pulsating organisms, the continuum whose flow we persist against. And so we invent time.

It is through resistance and peristence that we are, paradoxically, conscious of time. As Laura Riding austerely put it: 'as thinking beings we have greater resistance to change than material things have; our greater resistance gives us time to reflect on our position.' This is in effect an intellectual orientation – knowing where we stand. Seymour-Smith has pointed out that Riding was given to restating early Christian 'gnostic' beliefs. Riding's statement is a gnostic one: we live through knowing. (In Gnostic terms the immaterial and immortal Light of our thinking opposes the materiality and mortality of Darkness.) But reflection can exist at a much more primitive level than ours. Even a jellyfish, although like an amoeba it is probably 'part of its own hypothesis' shows 'reflection' in its behaviour although we don't know what it thinks. A jellyfish is conscious; a stone is not. A jellyfish pulsates; the stone merely vibrates. Whether entrained with others or alone, the jellyfish orients itself to each Now.

Ultimately persistence is in the face of death. ('Beharrung' was much valued by German army generals during World War II). As the American poet Trumbull Stickney wrote, 'He stubborned like a mighty slaughter beast… He has the fright of time.'

The duality of the universe

I think the duality reflected by the two main competing cosmological theories of today, the Time Universe and the Shape Universe is a reflection of our own duality. The Multiverse theory, bizarre as it first appears, represents an attempt to reconcile the two halves of the duality by stating that not only can they exist at the same instant – i.e. we can live in both universes – we can live in all sorts of universes.

All three universes contain life. The ultimate duality is not in us, however, it is 'out there'. It is the duality between the living (pulsation) and the non-living (pulse-waves). We experience this duality in ourselves. And wherever the living exists, it experiences this duality.

To restate the evidence described earlier: the series and event clusters of life *persist against time*. This is more than a restatement in parallel terms of the ultimately pessimistic view that life is negative entropy. But it can be expressed as a view of two opposing forces (or perhaps directions, if energy is left out): *time* and *against time*. Whether there is any relation between these forces and the conjectured opposition between *matter* and *anti matter* (reminiscent of the Gnostic Dark versus Light) remains to be seen. Our experience is *time / no time*.

Barbour's universe consists (as described in a paper on relational dynamics, 2001) of a 'unity of unities'. The phrase is from Giordano Bruno. It was from reading (or reading of) Bruno that Reich modified his 'life energy' to 'the cosmic energy ocean'. Before I had read this phrase of Reich's, in 1970, I published a book of poems called *The Ocean Everywhere*. They originated in my sense of pulsation throughout nature and in the atmosphere. At the time I had not distinguished pulsation from pulse-waves. But the poems express also a sense of resistance to the ocean everywhere and distinction from it. If we *are* the ocean everywhere, we are annihilated in our unity with it. Perhaps, as Reich speculated with newfound religiosity, we return to the cosmic ocean at death. In Buddhism this is called returning to the divine ground. I prefer to think we exist within the 'ocean everywhere' but as distinct beings. This is no doubt why Leibniz's version of the universe appeals to me.

Barbour's version of the universe is 'Leibnizian' as well as Machian. But in a recent paper with Niall O'Murchadha on *Conformational Geometrodyanmics* this universe takes on a dual aspect. Barbour and

O'Murchadha suggest that there are two opposing or balancing forces in the universe: gravity and a non-gravitational force – but not 'anti-gravity', which has also been conjectured. They explain Hubble's red shift as other than an effect of expansion. They propose a new 'cosmological force' which according to their calculations cancels out the effects of the known forces in fields (light, electromagnetism etc.). The only other force that remains is gravity. Gravitational force is weak and there are great problems in its measurement, but Barbour and O Murchadha hope that the inclusion of the cosmological force as a constant will enable the slight changes detected in the new super powerful gravitational measuring chambers to be mathematically expressed. Barbour proposes that the cosmological force (he does not name it) is like a blue shift to counterbalance the red shift. Barbour and O Murchadha are describing a *dual universe*.

Sex

Back to Parmenides, one last time. *Peri Physeos* includes a fragment, at the end, which is sometimes claimed to be a later addition because it suddenly introduces a new theme – that of the determination of male and female sex.

> When man and woman mingle the seeds of love that spring from their veins, a formative power, maintaining proper proportions, moulds well-formed bodies from this diverse blood. If, when the seed is mingled, the forces clash and do not fuse into one, then cruelly will they plague the offspring with a double-gender. On the right [of the womb] boys, on the left girls.

If the duality of Parmenides' world view is accepted – the 'doubleness' of the vision articulated by the goddess, the co-existence of the rushing journey through life experienced by the male narrator and the static plenum of unity experienced by the female goddess – then the theme of the doubleness of the foetus in the womb is relevant. As previously noted, Parmenides' comment on mixed sexuality only now comes into its own with the evidence from neuroscience that in some cases of disturbances of the usual hormonal environment of the womb a foetus may be 'programmed' to have a male body and a female brain (and vice versa), the programme activated (the film developed, as it were, through exposure) by adolescent development and experience. Leaving aside the question of the ambi-sexuality (a more accurate term, as Seymour-Smith has pointed out, than homosexuality or bisexuality) which characterises between 1.5% (according to a huge 2010 survey in the UK) and 4% (a similar 2012 survey in the USA) of human beings, it is established that gender depends on whether the ovum has been fertilised by an X or Y spermatozoon (in itself no more alive than a virus but it contributes to life once it is enveloped in a cell, the ovum). However, even if the ovum is fertilised by a Y spermatozoon it will not develop into a male foetus unless the 'right' kind of biochemical environment is provided in the womb. Its initial tendency is to be female. Males are only new versions of females, differentiated out of the basic female 'ground' of the embryo.

Parmenides could know nothing of these details. But he must have known that at a certain stage of development the foetus becomes visibly sexed. His fragment adds further evidence to his depiction of a dual universe. And he is proposing that the sex of a person is determined by his or her *position* in the womb – on the right or on the left. We know this is wrong. But the principle is consistent with the goddess's vision of a plenum. In a timeless plenum – as in Barbour's 'Platonia' or Mach's universe of inter-relations – what counts for your identity is literally *where you are*.

The word 'sex' as we now use it originates in a poem by John Donne in which he writes:

> ….should she
> Be more than woman, shee would get above
> All thought of sexe, and think to move
> My heart to study her, and not to love.

The Indo-European roots of the word 'sex' are in words meaning 'cut off'. This meaning survives in '*scis*sors.' Donne is pointing out (paradoxically – he loved paradox) that 'sex' means a putting together, in love, of what is cut off into two sexes. If his woman friend thinks she is more than woman, and above her own sex, then this isolates her from him and instead of loving her he can only study her.

In our human lives the most obvious duality is that of sex. It would be tempting to add it to the basic textbook definitions of life. However it is not included for the reason that if it were, then bacteria that reproduce by fission and demonstrate no sexual differences would not count as alive, yet they demonstrate the other attributes of life – reactivity, reproduction etc. Bacteria, technically 'prokaryotes', have their place in the evolutionary 'tree of life', starting with unicells and ascending to humans. Viruses are excluded, though 'life-like' in some ways, as they are seen simply as information packages, like genes, which cannot exist without being inserted in living cells. But if bacteria and viruses are both, as at least some evidence suggests, breakdown products of life which cause disease through 'defection' as well as infection, then bacteria's position in the tree of life is less secure. It can be argued that all forms of life are interdependent, but viruses and bacteria are totally dependent on cells and originate in their breakdown, and die without them. Using pulsation as the key criterion of life, the jury is still out. Viruses are too small to be observed under the light microscope and can only be examined dead. Bacteria do not appear to pulsate under the light microscope – but then nor would other hard-shelled organisms such as beetles: it is the insides of a beetle that pulsate. Under the light microscope it is impossible to see the various hairy-looking protrusions and propellers on the bacterium surface (these are visible when frozen in the electron microscope) and to determine whether or not these show pulsatory movement, as similar protrusions do in unicells such as vorticellae.

Unicells, protists (formerly called protozoa), technically 'eukaryotes', do pulsate, both in the heart-like 'contractile vacuole' and at the leading edge of the 'cyclosis' or 'cytoplasmic streamings' around the inside of their membranes as they move along. Whether they are 'sexed' is not clear. They reproduce in several ways: binary fission, multiple fission, and conjugation (joining and exchanging cell contents). Intensified pulsatory movement of the cytoplasm has been reported in observations of protists during 'conjugation'.

Flowers are sexed, whether in the same flower or in different kinds of flowers on the same plant, as in pear trees. Curiously, trees with blossoms of both sexes don't commit incest with themselves, as it were. A pear tree cannot bear fruit unless there is another pear tree within reach of the local bees, so that the male flowers of one tree can fertilise the female flowers of the other. In some cases – for example chestnut trees where most produce white 'female' blossoms but a few pink 'male' blossoms – whole trees can be described as sexed.

Whatever the biology or botany the theme is constant: the coming together and unification of two distinct forms which have been separate and distinct leads to reproduction. Conversely, there is no evidence whatsoever of sex in the non-living such as the mineral world of rocks and lava, or in the planets or the sun or the stars. This morning I sat having breakfast with my wife in the conservatory looking out onto our back garden. The primary fact of our life together is that we are a man and a woman. We are obviously two distinct types of human being. But the table we eat at is not sexed. My knife and fork are different and complimentary but they are not sexed. The paving slabs outside the conservatory door are not sexed.

To all intents and purposes, the most conspicuous fact of our being humanly alive is our sex. It is the initial given. 'Is it a girl or is it a boy?' And the sexual difference, although most obvious in the shapes of our bodies, also exists inside our bodies where every cell of a male body is distinctly male and every cell of a female body is distinctly female. (This continues to apply in ambisexuality or trans-sexuality). Furthermore, there are distinctions between the male and female brains, not huge, but throughout. And many people would claim from experience of themselves and others that there is a basic difference between the male and the female *minds*.

This is not a difference in intelligence or ability. Neuropsychological tests show that there are no significant differences between male and female groups (of the same population) in general intellectual abilities. I have also concluded in my book *Emotional First Aid* that there is no evidence of differences in the capacity for emotional expression. For example, 'big boys don't cry' is an observation of social conditioning, not biology.

But the bodily experience of being a woman is different from the bodily experience of being a man. And the mind is not, I think, separable from the body. Furthermore mind and memory are one. The most articulate 20th century proponent of the idea that men and women have distinct minds was the poet Laura Riding. At one point she rejected 'the damned thing' – sex. But even then she thought men needed women's *minds* to connect them with 'unity'. Unfortunately she demarcated mind from body, to the extent that in her study of language, *Rational Meaning* (written jointly with her husband Schuyler Jackson), she aims at a 'voiceless language' – i.e. a disembodied abstraction without the physicality of sounds attached. The Jacksons want to disembody language in their insistence that meaning resists change. But the mind/ memory of each sex includes their bodily sensations as well as their thoughts. Poetry is feeling thought. Metaphor, which has such a huge input in language that it creates much of any vocabulary, is (as Julian Jaynes first pointed out) rooted in bodily sensations.

The time sense

Although Laura Riding and Wyndham Lewis were at loggerheads on most issues, both disdained what Lewis had labelled the 20th century time cult. This cult did not exactly worship time, but as Lewis pointed out it was obsessed with time, either in following it mindlessly or in trying to subvert it. An example was the stream-of-consciousness writing first made famous by James Joyce. This mimics the flow of time but at the same time jumbles time sequence. This was taken as realism, but in fact most human consciousness is well organised in time terms and thinking follows time sequence. Only in acute confusion or delirium (Falstaff babbling of green fields) is time jumbled. Riding called stream-of-consciousness writing 'a disorder of the time sense'. There *is* a time sense, almost always with us when we are awake and even in dreams. It exists at the interface of our individual pulsation and pulse-waves at large. And without this time sense we cannot have the opposite sense of being 'against time'. Only as we are aware of the time sequence we live in can we be aware of the enduring meanings we set against it.

So many love poems are about whatever endures against time, or as the Welsh poet Alun Lewis put it, 'the question of what survives of the beloved', that clearly the intensity of the sexual relationship – meaning not just in the body but in the minds – brings out thoughts about its end, whether in the sense of orgasm as an end, or death and separation as ends. Perhaps, as Charles Sorley wrote 'to poets' just before he was killed in the First World War, 'Ours the exceeding bitter agony / Yours the exceeding bitter cry'. The man woman relationship – like all the human dualities – intensifies the sense of time.

The ocean everywhere

If the fundamental duality in the universe is between life and non-life, then in spatial terms we amount to nothing – mere specks – in the vastness of the non-living universe. (Unless Giordano Bruno was right and this is an infinite universe with living beings almost everywhere – a vision not unlike Deutsch's multiverse). In a sort of existential despair we resort to inventing God or gods or the Laws of Physics which make us feel less alone. But as David Stove remarks, we are 'top dog' in the universe we know. The possibility that we are the most complex things in the universe is not necessarily a consolation. We are as much flawed and destructive as we are beautiful and creative. Or *almost* as much. As Robert Frost remarked, the world must be at least 51% all right or we would have given up and become extinct long ago. And even the vastness of space may be an illusion. Barbour and his colleagues, or perhaps their eventual successors, may demonstrate that the universe is a plenum in which *space* is an illusion as well as time. The Space universe would become a Place universe! Of course we would still occupy much less mass in the universe than the lifeless rest. But our aliveness is in our complexity and, as noted above, perhaps we contribute more to the *relatedness* of a spaceless, timeless universe than we think we do. For one thing we contribute time itself. It is intrinsic to us even as we live 'against' it.

There is a difference between *duality* and *division*. Non-life is full of examples of division – for example the split flow of a river around a rock, or the continual splitting of the ink drop falling down through the water. It also contains examples of division and unity or re-unity – for example the division of the aurora borealis into two streams of pulsing light which then 'couple', the light intensifying, and separate again. No wonder Reich compared this 'cosmic superimposition' with human sexual superimposition and orgasm. It is tempting to seek further and further unity in the universe – the 'implicate universe' of some quantum physicists, the 'divine ground' of Buddhists, the 'unity of unities' of Leibniz. As individuals (un-dividables) we have been born, separated from the ground and unity of our mothers. Literature and psychoanalysis provide examples of the longing, men's particularly, to return to the womb. The psychoanalyst Sandor Ferenczi wrote in Thalassa of the longing to return to the womb and to the cosmic ocean. On the other side, there is an early Irish poem attributed to St Ita where she imagines herself breast-feeding baby Jesus.

No doubt we, the living, are subject to the same physical laws as 'the ocean everywhere'. In Shape Universe terms we presumably include the same patterns, or almost the same. The universe seems to contain near but not exact correspondences which can be seen in terms of almost corresponding shapes. An example is the levels of correspondence between the sounds of a piece of music, the pattern of its notes written out in a score, and its possible expression in mathematical notation. The observation that musical intervals and mathematics correspond is said to originate with Pythagoras and it led to centuries of assumptions about the 'heavenly harmony' of the universe. But in the 18th century it was discovered that the correspondences are not exact. As Leibniz pointed out it is not possible that any entity in the universe is identical to another. If it were identical it would *be* that other. And it would be in that other's place. Each entity (monad) has its own place in the universe. Although we see similarities and correspondences and symmetries, ultimately every thing in the universe is a singularity.

Hence Barbour's Shape universe is a vast configuration of configurations where, as I understand it, nothing duplicates itself. The relations through 'best matching' that keep the universe together (with no need for time or space) are endless 'Nows' among countless points each of which is distinct from the other. I would add that some of these points are related to each other in configurations that contain pulsation and that these configurations are alive and aware, the remaining configurations not. But I do find this whole language of 'configurations' very abstract. And Barbour does not seem able to fit life or the 'mystery' of consciousness into the graveyard of 'Platonia' where time exists merely as an illusion structured into 'time capsules.'

Perhaps in this context the 'event clusters' identified by Kammerer and evident in more-than-coincidence and the genesis of poems can be seen as a form of best matching. We may experience subjectively the patterns that hold the universe together.

Barbour's is a static universe and therefore unavoidably feels dead. Yet he acknowledges (though indirectly as he states that time exists to prevent everything from happening at once) that in this static universe everything must be happening at once. This is the paradox of Parmenides again.

Language, flux, and meaning

The linguistician Alfred Korzybski proposed that language should be changed to make it consistent with science. Hence (as Jackson and Jackson explain in *Rational Meaning*):

> The term "space-time" is an example of Korzybski's idea of the kind of improvement he thought could be made in the structure of language. Regarding the term as scientifically right, he welcomed it as a replacement of "time" and "space": "time" and "space" were for him by their meanings wrong words, words that imposed a "delusional verbal split" on language and made it accord thus with a primitively fictive world-structure.

After the 20th century we should know the dangers of what Orwell called 'Newspeak' – attempts to impose a new language in order to ensure obedient rigidity of thought, now known as 'political correctness.' Most of us resist the sort of linguistic hygiene which wants to eliminate a 'delusional verbal split'. Language (as Jackson and Jackson emphasise) does not have to subordinate itself to changes suggested by science which occupies an infinitely smaller corner of the truth than language. The words 'time' and 'space' have their meanings. As does the new word 'space-time'. 'Space-time' does not need to put its parents 'time' and 'space' against the wall and shoot them, or even to betray them to the authorities. If a new physics concludes that space and time and therefore space-time are non-existent, whatever new words replace them need not eliminate them. Space and time will exist – being structured along with other dualities into our bodies and thus our minds – even as we discover new ways of naming our experience. This very naming of experience is part of our being alive 'against time.'

Language itself, too big a subject to be included in this book, incarnates Parmenides' paradox. Twentieth century linguistics, swept up in the time cult, has emphasised change in language – i.e. in usage, vocabulary, and pronunciation. Language includes the Heraclitan flux. But it also *contains* it, in the form of meaning, and meaning endures. Although not quite static the meaning of each word is consistent. As Shakespeare said, 'Truth is truth until the end of reckoning.' Leaving aside particular voices and associations, the word 'truth' as I speak it is identical in its meaning to the word 'truth' as you speak it. In language Leibniz's theory breaks down: words *are* identical in their shared meaning. But the word 'truth' is not identical in its meaning (or its form, obviously) with the word 'truthfulness.' Their distinct meanings resist the linguistic suicide bombers of the time cult who urge the total relativeness of meanings according to context and usage, so that 'anything goes.' In Alice in Wonderland, Humpty Dumpty says 'When *I* use a word… it means just what I choose it to mean – neither more nor less.' His fate is well known. But I have met university graduates who refuse to clarify word meanings on exactly Humpy Dumpty's grounds. If flux rules in language, then sense falls apart. As Robert Graves wrote:

> There's a cool web of language winds us in…
> But if we let our tongues lose self-possession,
> Throwing off language and its watery clasp
> Before our death, instead of when death comes,
> Facing the wide glare of the children's day,
> Facing the rose, the dark sky and the drums
> We shall go mad no doubt and die that way.

Change as time

There does seem to be a broad duality that permeates the entire universe – as expressed by the Chinese Yin / Yang diagram which is reminiscent of coupling auroral streams. But Yin / Yang is not an either / or. Each is simultaneously present, as in Leibniz's Unity of Unities, Goethe's Unity of Ones, and the archer Howard Hill's 'split vision.'

Yin / Yang although it works itself out at the human level where Yin is the female, Yang the male, is universal – a sort of cosmic plus and minus. It even influenced the development of calculus by Leibniz who proposed that the Yin and Yang hexagrams of the *I Ching* formed a binary arithmetic. *I Ching* is usually translated as 'Book of Changes' but it literally means 'Change Persistence'. This seems to represent a nice ambiguity of 'persistence of change' and 'persistence versus change' – the theme of the present book – and to make them simultaneous: Time / No Time.

The I Ching is said to have been invented by Fu Xi no less than 5,000 years BC, but its Taoist and Confucianist accretions have led it so far from its origins that it is difficult to return to its essentials. These seem, though, to have been the representation of the Yang (the male) and the Yin (the female) as respectively an unbroken and a broken line (in effect a penis and a vagina). Chinese archeologists have found patterns of six lines (hexagrams) engraved on stone and on tortoise shells from thousands of years BC. But these were solid lines. Eventually broken lines appeared, combined with the unbroken as Yin and Yang respectively into patterns of 3 lines – trigrams. If two lines represent simply 'man' and 'woman', it is impossible to be sure what the third line – which expands the dynamics of the trigram – originally represented. Perhaps simply another man or woman. The greatest commentator on the *I Ching*, Wang Bi (c.250 AD), as a modern editor attests, 'describes the actions and interactions of the lines as if they were people.' But the history is of a gradual abstraction of the Yin and Yang trigrams, starting from pairs of opposites indicating the main features of nature: heaven / earth, fire / water, thunder / wind, mountain / marsh. Heaven (Yang) / Earth (Yin) becomes Force / Field, then Creative / Receptive, then Dominant / Submissive and so on, until the Confucian code of the dominant man and the submissive woman becomes enshrined in the system. The I Ching as it exists now is influenced by social beliefs more than biological observation. But as Wang Bi put it, 'The hexagrams deal with the moments of time, and the lines are concerned with the states of change that are appropriate to those times.' The ancient Chinese did not separate time from change. The original I Ching defines the recurring shapes in what was in effect a Shape universe. If one shape is man and the other Woman, perhaps the third line of the trigram expresses them as One.

Fritjof Capra's *The Tao of Physics* claimed to be based in the Chinese Tao – the 'Way', a quasi-religion developed long after the original I Ching. Capra proposed a total animism: everything is alive. This meets what

David Stove called 'the basic demand of religion: the universe is indeed our kindred.' In the universe where everything is alive this includes even the specks of dust our corpses become. We never really die, we just become another level of being – or whatever the mystical mind dreams up.

The timeless universe, whether in the Deutsch or Barbour version, also has its consolations. At this moment one Seán is typing these words and another is making a cup of coffee or perhaps doing something even more interesting. Or the Seán of 69 typing these words is simultaneously 6 years old and picking bluebells with his family on Cave Hill, or 16 years old and taking a walk with his girlfriend. Or I am still being born – *over there* in 1943.

But the other side of the everything-is-alive coin is that everything-is-dead. There is no difference. This death trap is hard to avoid in visions of the timeless universe. Barbour frets over the 'graveyard' of Platonia.

Even in the Shape universe our beginning and our end are different experiences. I am a realist, not an idealist who believes the universe consists of my experience of it. The dualism of life and non-life may exclude the idea of the non-living universe as 'kindred.' It's pretty cold out there. But we *are* alive in the simultaneous nows of change – and we, the living, experience change as time. Through time we live, and through time we die.

Eternity in an hour

The life / non-life duality has not appeared in Western physics since its first expression in Parmenides' *Peri Physeos* – and even there only one side of the goddess's dual vision was recognised by Plato and his successors in idealistic philosophy. None of the three theories of the universe discussed earlier have any place for dualism. Current explanations in physics leave dualistic phenomena unexplained. To return to my own experience at school in the 1950s where physics was taught as if the quantum revolution and even relativity had not occurred, a further element in my disillusion with science was its lifelessness. It consisted merely of learning formulas and performing chemical experiments with various minerals. (Spatialisation and Mathematicisation are *dead*.) By contrast, at my first school we had done 'Nature Study'. At the age of 8 or so I was bringing to school soil samples dug carefully from my garden to show the layers of organic life, and I was also studying the stars at night and bringing in drawings of constellations – I remember one of Bootes rising over Cave Hill below the tail of the Plough. The dualism between the night sky and the earth was apparent even then in the different feelings I remember they evoked: a kind of awe for the night sky and a more intimate curiosity for the earth.

How fantastic it would be if the universe consisted of the interaction of separating and unifying principles, as in sex. As I observe the aurora I am *moved* by it and I see the similarity between my own 'streamings' and couplings and the aurora's. Perhaps biological sexual division is part of a universal division – Yin / Yang. But we are nevertheless distinct. The auroral sub-storm is not an orgasm. The auroral streams pulse, but we pulsate. Our awareness has developed, perhaps, out of resistance to the cosmic flow that it also creates. When our awareness is at its most intense – in orgasm or in a poem, for example – we feel we are no longer part of the flow of time. It has widened out in us to the point where we are ourselves – or ourselves and our other – in our own place in the universe.

I have emphasised the distinction between pulsation and pulse waves, and after summarising three theories of the universe I have proposed a dual universe, the duality being life and non-life. From this duality, other dualities follow, on the life side: the duality of sex, and the various dualities in the structure of our brains. Years ago, in the Quebec midnight when I stood observing the coupling of auroral streams, I had the sense that this coupling was inhuman, not the same as the coupling between man and woman, but that each reflected the other. The living and the non-living are distinct, but they are both subject to the recurring shapes of change.

The final paradox is that we have got time the wrong way round. We think of time – and measure it – *out there*, in the universe. But the universe is timeless. Our clock measurements are simply lengths. Time is not out there it is *in here* – in us. We invent it as we live it. And our days are numbered by the external clocks we identify all around us. Yet we also experience, in occasional discontinuities, the timelessness of the universe of which we are part. Although we die – we are finite – we are eternal too. As William Blake wrote:

To see a world in a grain of sand
And heaven in a wild flower,
Hold infinity in the palm of your hand
And eternity in an hour.

The Unshaking Heart of Truth

In this explanation of the paradox of poetry and physics, there is one thing left out: value. If we settle for the relational universe (or universes) of modern physics, where is what Parmenides called 'the unshaking heart of truth?' Isn't *everything* true – relatively of course – in a relative universe? What makes a good poem (or painting, or piece of music, or scientific theory) different from a bad one? By the early twenty-first century the 'anything goes' relativism of a century before has become so pervasive that it is difficult to write such words as good and bad without immunising them with quotation marks: 'good' and 'bad.' As for the longstanding connection between good and true – as when a carpenter inspects a true (rather than crooked and therefore more likely to break) grain in a piece of wood and says 'she's good' – forget it!

This book has been about time and paradox, not about values and ethics. But in writing about poems, at least, I have allowed myself to imply that the inspired poem emerging spontaneously from a chain of more-than-coincidence is better and more true than a concocted and factitious word picture or tract. My old friend Martin Seymour-Smith used to emphasise that the poet has to *face* himself or herself in a poem. What does this mean?

It is what that extraordinary poet/ physicist Parmenides sums up as – or the Goddess in his poem sums up as – 'the unshaking heart of persuasive truth.' Again and again the Goddess admonishes the traveller to think only of what *is* – never of what is *not*; ignore the ideas of others; 'the same thing is there for thinking and being'; 'gaze even on absent things with a present mind, and do so steadily.' This mindfulness, this extreme focused attention to what is present (which of course includes the past, since as neuropsychology now demonstrates, our perception of the present is already of the past) is 'good' as distinct from the mental slackness, distraction by the ideas of others, evasion of the truth, *not facing* either oneself or the world which is in effect 'bad.' Parmenides does not use the words good and bad, but his vision includes the Goddess's exhortations to a certain kind of behaviour. Values and ethics are intrinsic to *Peri Physeos* as to any other true poem. It is difficult to face oneself, or the truth. It may be painful. It requires so much attention. It is *in*tensive, rather than *ex*tensive.

We *value* certain poems, or pieces of music – or for that matter certain qualities and events in a human relationship which turn it to love – to the extent that they become close to the unshaking heart of truth. And how do we recognise the truth? It *moves* us, and in that moment we both stand outside time and create it.

BIBLIOGRAPHY

Akasofu, S.I., 1981. 'The Aurora'. *American Scientist*, 69, 5.
Alfven, H., 1963. Cosmical Electrodynamics. Clarendon Press, Oxford.
   1975. Evolution and Structure of the Solar System. Reidel, Boston.
Austin, S., 1986. Parmenides – Being, Bounds and Logic. Yale, Newhaven.

Bailey, W.R. & Scott, E.G., 1974. Diagnostic Microbiology.
Bak, P., 1996. How Nature Works. Springer.
Baker, J.W. & Allen, G.E., 1975. Matter, Energy and Life. Addison & Wesley, New York.
Barbour, J.B. & O'Murhcadha, N., 2002. Conformational Geometrodynamics.
Barbour, J.B., 1989. 'Maximal Variety as a New Foundational Principle of Physics' in *Foundations of Physics*, 19, vol.9.
   1992. 'On the Origin of Structure in the Universe.' (unpublished paper).
   2000. The End of Time. OUP, New York, London.
   2001. The Discovery of Dynamics. OUP, New York.
Baross, J.A., 'Geomicrobiology of Hydrothermal Vents', 2003, Aquatic Science Meeting.
Barrett, E.C., 1974. Climatology from Satellites. Methuen, London.
Battan, L.J., 1974. Weather. Prentice Hall, New Jersey.
Bergson, H., 1959. Oeuvres. (Works). Presses Universitaires de France. Paris.
Bessis, M., 1973. Living Blood Cells. Springer Verlag.
Birkeland Symposium on Aurora and Airglow. Sandefjord, Norway, 1967.
Boadella, D., 1973. Wilhelm Reich: The Evolution of his Work. Regnery, Chicago.
Bohm, D., 1995. Wholeness and the Implicate Order. Routledge, London
Bortoft, H., 2004. The Wholeness of Nature, Goethe's Way of Science. Floris, Edinburgh

Brody, H., 1982. Maps and Dreams. Pantheon, New York.
Buck, C.D., 1949. A Dictionary of Selected Synonyms in the Principal Indo-European Languages. University of Chicago Press.
Buñuel, L., 1983. My Last Sigh. Knopf, New York.
Burnham, D., Gottfried Wilhelm Leibniz. Internet Encyclopedia of Philosophy.
Burr, H.S., 1972. Blueprint for Immortality. Spearman, London.
Butler, H., 1985. Escape from the Anthill. Lilliput, Dublin.
Butler, S., 1878. Life and Habit. Trubner, London.
    1910. Unconscious Memory. Fifield, London.
Buvat, R., 1969. Plant Cells. World University Library, New York.

Capra, F., 1975. The Tao of Physics. Random House, New York
Carroll, L., 1896. Through the Looking Glass. London.
Chamberlain, J.W., 1961. Physics of the Aurora and Airglow. Academic Press, New York.
Chargaff, E., 1978. Heraclitean Fire. Rockefeller University Press, New York.
Coleridge, S., 1817. Bibliographia Literaria.
Coxon, A.H., 1986. The Fragments of Parmenides. Van Orcum, Netherlands.
Cozolino, L., 2002. The Neuroscience of Psychotherapy. Norton, New York.
Cytowic, R., 1996. The Neurological Side of Neuropsychology. MIT Press, Cambridge, Mass.

Damasio, A., 2000. The Feeling of What Happens. Vintage Books, New York.
Dawkins, R., 1976. The Selfish Gene. OUP, Oxford.
Desmond, A. & Moore, J., 1992. Darwin. Penguin, London.
Deutsch, D., 1997. The Fabric of Reality. Penguin, London.
Donne, J., 1611. An Anatomie of the World. London.
    1995. Complete Poetry. Nonesuch, London.
Dunne, J.W., 1929. An Experiment with Time. Black, London.

Encyclopedia Britannica, 15th Edition, 1981.

Ferenczi, S., 1968. Thalassa. Norton, New York.
Fisher, S., 1973. The Female Orgasm. Basic Books, New York.
Frost, R., 1930. Collected Poems. Holt, New York

Gimbutas, M., 1982. The Goddesses and Gods of Old Europe. Thames & Hudson, London.
Goldberg, E., 2009. The New Executive Brain. OUP, New York.
Gough, S., 2012. The White Goddess: An Encounter. Galley Beggar, Norfolk.
Grad, B., in *International Journal of Parapsychology* 3, 1961; 5, 1963; 6, 1964.
    in *Journal of the American Society for Psychic Research* 59, 1965; 61, 1967.
Graves, R., 1940. The Long Weekend. London.
    1948. The White Goddess. Faber & Faber, London.
    1967. Poetic Craft and Principle. Cassell, London.
    2000. Complete Poems. Carcanet, Manchester.
Gregory, R.L., 1973. Eye and Brain. OUP, London

Haldane, S., 1977. Human Pulsation, PhD dissertation, Saybrook Institute, San Francisco.
    1984. Emotional First Aid. Station Hill Press, New York.
    1992. Desire in Belfast. Blackstaff Press, Belfast.
    1996. John Donne. Greenwich Exchange, London.
    1999. Thomas Hardy. Greenwich Exchange, London.
    2000. Lines from the Stone Age. Greewich Exchange, London.
    2009. Always Two. Collected Poems 1966-2008. Greenwich Exchange, London.
Hall, T., 1969. Ideas of Life and Matter. Chicago and London.
Hardy, T., 1922. Late Lyrics and Earlier. Macmillan, London
Harrison, J., 1921. A Prolegonema to the Study of Greek Myth. Macmillan, London.

Harvey, W. 1628 De Motu Cordis et Sanguinis (On the Circulation of the Heart and Blood).

Hawking, S., 1988. A Brief History of Time. Bantam Press, London.

Hill, H., 1953. Hunting the Hard Way. The Derrydale Press, Lanham, Maryland.

Ho, M-W, 1998. The Rainbow and the Worm – The Physics of Organisms. World Scientific Press, London.

Hobson, J Allan., 2002. Dreaming. An Introduction to the Science of Sleep. OUP, Oxford.

Housman, A.E., 1933. The Name and Nature of Poetry. Cambridge University Press.

I Ching (as interpreted by Wang Bi), ed. Lynn, Richard. 1994. Columbia University Press, NY.

Inglis, B., 1981. The Diseases of Civilization. Hodder & Staughton, London.

Internet Encyclopedia of Physics, 2006

Jacob, F., 1973. The Logic of Living Systems. Allen Lane, London.

Jackson L (Riding) & Jackson, S., 1977. Rational Meaning. University of Virginia Press.

Jaynes, J., 1977. The Origin of Consciousness. Houghton Mifflin, Boston.

Jezzard, P., Matthews, P.M., Smith, S.M. Functional Magnetic Resonance Imaging: An Introduction to Methods. Oxford University Press, 2001.

Kammerer, P., 1919. Das Gesetz der Serie. Deutsche Verlags-Anstalt, Berlin.

Kammerer, P., 1922. Allgemeine Biologie. Berlin.

Kilner, W.J., 1911. The Human Atmosphere. London.

Koestler, A., 1971. The Case of the Midwife Toad. Hutchinson, London

Koestler, A.,1972. The Roots of Coincidence. Hutchinson, London.

Kraus, F., 1926. Allgemeine und Spezielle Pathologie der Person. Leipzig.

Krippner, S & Rubin, D., 1974. The Kirlian Aura. Anchor /Doubleday, New York.

Krippner, S. & Ullman, M., 1973. Dream Telepathy. Macmillan, New York.

La Mettrie, O., 1960. Man the Machine, ed. Vartanian. Princeton, N.J.
Lehmann, W., 1999.Gesammelte Werke Vol.8. Klett-Cotta, Stuttgart.
Levi-Strauss, C., 1966. The Savage Mind.
Lewis, A., 1945. Ha! Ha! Among the Trumpets. Allen & Unwin, London.
Lewis, W., 1927. Time and Western Man. Chatto & Windus, London.
Loftus, E., 1980. Memory. Addison Wesley, New York.
Lovelock, J., 1979. Gaia. OUP, London.
Luce, G.G., 1971. Biological Rhythms in Human and Animal Physiology. Dover,
Luria, A., 1967. The Working Brain. Penguin Books, London.
    1925. 'Psychoanalysis as a System of Monistic Psychology'. Moscow.

Mach, E., 1922. Die Analyse der Empfindungen. Fischerverlag, Jena.
MacLean, S., 1999. As Choille gu Bearradh – Collected Poems. Carcanet, Manchester.
Manwell, R.T., 1968. Introduction to Protozoology. Dover, New York.
McCormac, B. (ed.) 1967.Aurora and Airglow. Reinhold, New York.
McGinn, C., 1992. The Problem of Consciousness. Blackwell, Oxford.
    1997. Ethics, Evil and Fiction. Oxford University Press, London.
Medawar, P., 1967. The Art of the Soluble. Methuen, London.
Monod, J., 1970. Le Hazard et La Nécessité (Chance and Necessity). Ed. Seuil, Paris.
Moss, T., 1979. The Body Electric. Tarcher, Los Angeles.

Orme, J.E., 1969. Time, Experience and Behaviour. Iliffe, London.
Owen, S., 1997. Anthology of Chinese Literature. Norton, New York.

Panksepp, J., 1998. Affective Neuroscience. OUP, New York and London.
    2002. Biological Psychiatry. Wiley, Chichester.
Persinger, M. & Lafreniere, G., 1977. Space-Time Transients. Chicago.
Pert, C., 1997. The Molecules of Emotion. Touchstone, New York.
Pierrakos, J., 1971. The Energy Field in Man and Nature. New York.

Popper, K. & Eccles, J., 1977. The Self and its Brain. Springer Verlag, Berlin.
Popper, K., 1962. Conjectures and Refutations. Routledge, London.
    1998 (posthumous). The World of Parmenides. Routledge, London & NY.
Portmann, A., 1967. Animal Forms and Patterns. Schocken, New York
Pribram, K., 1991. Brain and Perception. Lawrence Erlbaum, New Jersey.
Puharich, A., 1977. Beyond Telepathy. Vintage Books, New York.

Reich, W., 1964. Cosmic Superimposition. Farrar Straus & Girouz. New York.
    1961. The Function of the Orgasm. Farrar Straus & Giroux, New York.
    1973. Ether God and Devil. Farrar Straus & Giroux, New York.
    1979. The Bions. Farrar Straus & Giroux, New York.
    1987. The Bioelectrical Investigation of Sexuality and Anxiety. Farrar
Reichenbach, K von, 1977. The Mysterious Odic Force. Aquarian Press, London.
Riding, L., 1929. Contemporaries and Snobs. Cape, London.
Riding, L. et. al., 1937. Epilogue, III. Constable, London.
Ridley, Matt. 2003. Nature Via Nurture. HarperCollins, London.
Rivers, W.H., 1923. Conflict and Dreams. London.
Rosen, J. 2010. Lawless Universe. John Hopkins University Press, Baltimore.

Sanderson, I., 1972. Investigating the Unexplained. Prentice Hall, New Jersey.
Schroedinger, E., 1944. What is Life? London.
Seymour-Smith, M. 1975. Sex and Society.
Sheldrake, R., 1981. A New Science of Life. Blond & Braggs, London.
Smith, A. & Kenyon, D. Is Life Originating De Novo? (in *Perspectives in Biology and Medicine*, 15, 1972)
Smith, A., 'The Origin of Viruses' (in *Enzymologia*, 43, 1972)
Sorley, C., 1985. Collected Poems. Cecil Woolf, London.
Spilsbury, R., 1974. Providence Lost, A Critique of Darwinism. Oxford University Press, London.
Stapp, H., 1993. Mind, Matter and Quantum Physics. Springerverlag.
Stewart, I., 1998. Life's Other Secret. Penguin, London.

Stickney, J.T., 1903. Les Sentences dans la Poésie Grècque. Paris.
Stove, D.C., 1991. The Plato Cult and Other Philosophical Follies. Blackwell, Oxford.
   1995. Cricket and Republicanism. Quaker's Hill, Sydney.
   2006. Darwinian Fairytales. Encounter Books, New York

Thompson, D., 1942. On Growth and Form. Cambridge University Press.
Tomatis, A.A., 1963. L'Oreille et le Langage. Seuil, Paris.

Wheen, F, 2009. Strange Days Indeed. Fourth Estate, London.